Praise for *Science, Sι*

"'What am I? What is my purpo_ ___g. ᴅ.. ᴜ.. ansᴡers these timeless questions with evidence for the invisible subtle energies underpinning our physical world. The beauty of his approach is that understanding the science of subtle energies gives us a window into our true nature. Subtle energies are the élan vital, the vital force and pulse of life. They are the inherent creative principle responsible for human evolution and linked closely with consciousness.

"Dr. Gin takes us deeper into the potential of our creative principle with the freshness and aliveness of the truth [taught by] spiritual master Joachim Wippich. This book is rich with examples of his teachings. My favorite: 'Always create in Love, Harmony, and Balance.' This simple invitation is a gateway to creating our Golden Age.

"The practical tools presented herein lift the mystery of living and become self-evident: *we* are the source of our reality, conscious or not. This is good news, for we are in charge. We have intent, a powerful force that focuses subtle energy like a laser. Ascension is possible with an awakened and willing consciousness aligned with divine principles. 'I AM that I AM' is the basis of a new spirituality. Gin's work gives credence to Psalm 82:6: 'Ye are Gods.'"

"This book is a truly Amazingly, Delightful read!"

—Dr. Nisha J. Manek, MD, FACP; author, *Bridging Science and Spirit: The Genius of William A. Tiller's Physics and the Promise of Information Medicine*

"This unique book offers a vital vision for our times, reminding us of our miraculous Divine nature through the power of our thought, free will, and creation—which can make an enormous difference in our lives and the world, if only we would bring forth love, harmony, and balance. Dr. Jerry Gin's scientific and spiritual insight and wisdom inspires and shines through on every page. The book is packed with practical exercises and affirmations and invites the reader to greater empowerment and spiritual growth. It is easy to read and accessible to all."

—Beverly Rubik, Ph.D., Biophysicist; President, Institute for Frontier Science; Adjunct Faculty Member, California Institute for Human Science

"In today's hyperpolarized society, there is too often a tendency by many to close off from novelty. That could mean being closed off to novelty of experience; novelty of argument; novelty of risking being 'contaminated' by new information and data that might challenge our understanding about how we think the world works. These tendencies can manifest themselves in scientists as well as non-scientists.

"A traditionally trained scientist with a PhD in biochemistry as well as a successful entrepreneur, Jerry Gin has had the all-too-rare courage to be able to break away from traditional textbook thinking about how the world works when presented with evidence or experience that challenges conventional perspectives. This book is his extraordinary journey through novel worlds. Enjoy Jerry's journey. Hopefully, you will come away thinking about things in a different way. I know I did."

—BILL BENGSTON, PHD; President, Society for Scientific Exploration; author, *The Energy Cure*; Professor of Sociology, St Joseph's College

"The path for a new stage in human evolution is clearly explained by this exceptional book. Dr. Jerry Gin uses a unique approach to address the connection between science and spirituality, providing us with an opportunity to better understand the qualities of subtle energy as well as the essence of 'I AM.' He presents his conclusion first, and then provides scientific evidence and tools so that we can confirm the information for ourselves. The added bonus of affirmations by spiritual teacher Joachim Wippich offers the ability to raise consciousness, allowing us to help ourselves, the planet, and humanity."

—JAN WALSH, International Feng Shui Guild / BioGeometry Environmental Home Solutions Practitioner

"In this comprehensive and compelling work, Dr. Gin takes us on a journey through the energetic nature of reality to make the case that thought and intention are not only generative, but also measurable. Providing background scientific information and step-by-step 'how-to' instructions, this book shows you how to work with subtle energies for yourself and then asks the question, 'What does it all mean?' Drawing on the work of scientists and spiritual teachers alike, Dr. Gin gives us a glimpse into our true nature as creators, and challenges us to use our powers of thought and intention to create our world in love, harmony, and balance."

—JUDY KITT, President, Foundation for Mind-Being Research; Founder, Director, CEO, Life from the Ashes

"In Jerry Gin's previous book, *The Seeker and The Teacher of Light*, he described the historical journey that led him to discover many of the secrets of the universe, using the tools of both science and spirituality, and driven by his natural curiosity, by the incorporation of the 'coincidences' or 'synchronicities' that came his way, and especially by meeting Joachim Wippich, who opened his mind to better understanding the mysteries of the universe.

"This book approaches what he discovered (or uncovered) and applies these findings to a Science of Consciousness, which includes the realms of Subtle Energies, Science, and Spirituality as a Unified Field Theory. He starts from the top down by building the framework of his discoveries, then teaching readers how these discoveries can be useful to them in their own lives.

"As Tesla said, 'If you want to find the secrets of the universe, think in terms of *energy*, *frequency*, and *vibration*,' and the numbers 3, 6, and 9. With the additions that Jerry gives you in this book, you will be able to create your own tools to use your 'I AM' knowledge, then be able to measure your results and apply them to your own incarnation as a member of the 'I AM' Universe.

"If you want to see what Jerry discovered and how, read his first book; but if you want to cut to the chase, read this new book that shows you the answers he found, and how he (and then you) can apply them, and put them to the test for yourself. Get ready for a mind-expanding journey into areas you may have never before known existed. Enjoy!"

—MICHAEL BUCHELE, MD

"We are at a critical juncture in our humanity. Jerry Gin's new book gives us an incredible gift at a time when we most need it—the insight and wisdom to understand that we are not powerless, but are powerful creators of our reality. We are not divided; we are all connected. Gin combines history, science, and math with spirituality to support an investigation of energy, giving us the full picture of the essence of who we are. He shows us that we are not only surrounded by everyday magic and wonder, but in fact we are practitioners of it by our very nature. All the tools and resources we need to live well—to live fully in balance and harmony with one another and the Earth—are at our disposal."

—JANE MAJKIEWICZ, writeintuition.com

Science, Subtle Energies, and Spirituality

A PATH TO *I AM*

JERRY GIN

Torus Press

Torus Press
www.jerrygin.com
Email: *jerry@jerrygin.com*

Cover illustration:
Orgonite pendant designed by John Thompson, Mother Earth Orgone
(*http://www.mother-earth-orgone.com/*). Photo by Jerry Gin.

Developmental Editor / Editor / Publishing Coordinator:
Naomi Rose / *www.naomirose.net*

Typesetting and Book Design:
Margaret Copeland, Terragrafix / *www.terragrafix.com*

Printed in the United States of America
First printing 2022
ISBN# 978-1-7363982-2-7

ALSO BY JERRY GIN

*The Teacher and The Seeker of Light: On the Teachings of
Joachim Wippich and the Mystery of 3-6-9*. Torus Press, 2021.

The inspiration for this book remains
the wonderful teachings of Joachim Wippich,
the Teacher of Light referred to
in *The Seeker and The Teacher of Light.*
He is truly Amazingly Delightful.

Acknowledgments

My wonderful editor, who tries to make my writing simple to understand, is Naomi Rose. My text is better because she has reviewed it. My chapters become a book because of her.

My dear friends Jan Walsh and Rick Skalsky added their wonderful support. Jan put together the excellent compilation of Joachim's Affirmations in this book. She was very helpful in her review of content in the book.

The tools and brilliance of Ibrahim Karim, founder of BioGeometry, make the measurements of subtle energies possible, since he brought back to life the ancient learnings that had been forgotten. He is the discoverer of BG3, which forms a backbone to the work in this book.

My biggest supporter is my Amazingly Delightful wife, Peggy. She supports the many hours I spend on research, writing, exploring knowledge and understanding, as well as my work at FMBR and my companies.

Table of Contents

Introduction

Foundations of Science, Subtle Energies, and Spirituality

In this book, you will get to explore some fascinating ideas and realities about who you really are, what the world really is, what you can create with your thoughts, how healing can come about through ways you might not have considered, and much more. The basic, transformative premise is that you are one with the Creative Force in the universe—the I AM—and that, with this awareness, your experience of the world can be one of balance, peace, harmony, and love.

Because most of us have been raised within a scientific paradigm (involving premises, experiments, and an "I'll believe it when I see it" attitude), subtle energies (a major focus of what we'll be investigating) may be less familiar to you. Therefore—and because I am a scientist by training and inclination—let me begin our discussion with the *science* part.

The Science Part

Science typically deals with physical matter—the atoms and molecules that form our organs and bodies. According to that definition, our mind and consciousness reside in the brain. However, the field of quantum physics steps outside this belief system and acknowledges that we are "observers" who can affect outcomes of experiments by the simple presence of observing. Given quantum physics' view that reality is both wave and particle, we can "collapse the wave" and establish a reality that is based on our observations. Biology and chemistry, for the most part, do not think in those terms. Even quantum physics, although outside the Newtonian box of machine-like determination, tries to figure out how our brains change reality.

This book will introduce another concept as to our true nature, and provide scientific evidence that there is a different way to view who we are.

By applying principles of the scientific method to the study of subtle energies, we can receive valuable data and information as to the nature of who we are—in short, that our essence is I AM.

The Subtle Energies Part

What *are* subtle energies? Here is one definition:

> The concept "Subtle Energy" is often assumed to mean different forms of energy more subtle than conventional physical instruments can detect. In this sense, the concept comes from traditions that accept the human ability to see or feel forces that are not physically measurable. Subtle energy can also mean finding new understanding of how measurable forces produce effects; for example, processes in which electromagnetism can either produce profound changes with only tiny forces, or can register profound changes with only tiny forces.
>
> —Bernard O. Williams, Ph.D.,
> "Exploring Multiple Meanings of Subtle Energy"

The study of subtle energies offers tools to study our vibrational-energy levels. Understanding the information derived from studying *fields* helps us to formulate a sense of who we actually are. We know that magnets have a field, because it is the space where there is magnetic attraction or repulsion. Certain objects, certain states of matter, and magnets can create subtle energies in a space that are called *subtle energy fields*.

This book teaches how to detect and measure them. If we couple what was learned in *The Seeker and The Teacher of Light* with the information in this book, we come to know the I AM, which is the heart of our being.

The Spirituality Part

This book views *spirituality* in the light of the I AM. I AM is the essence of who we are, the energy and divinity that flows through all of us and everything. Once you understand the nature of creation, and can feel and test for your *own* vibrational energies, this point becomes self-evident.

This understanding will be supported by the scientific evidence for I AM, as shown in the experiments and the data described in this book.

When the understanding and embrace of I AM becomes the unifying thread of both science and spirituality, then humanity will reach a new stage in evolution. With the realization of who we really are, humanity will advance. Humans will create with the concept of harmony, balance, and love as the core principles of who we are. We will truly understand our connectedness.

The Purpose of This Book

This book has two primary purposes: to reveal our connectedness and thereby help humanity advance, and to provide you with tools for investigating the hidden world.

#1. *To Reveal Our Connectedness and Help Humanity Advance*

The primary purpose of this book is to present our true nature in the context of science, as viewed through (a) the study of subtle energies and our vibrational energies, as well as through (b) the teachings of Joachim Wippich (the "teacher" referred to in my previous book, *The Seeker and The Teacher of Light*), so that we can understand our connectedness and thereby help humanity advance. Our connectedness stems from our being all One. We are all God, experiencing all the ways of being. We are all I AM.

Usually, books begin at the beginning and present the conclusion at the end. In this book, however, I will present the conclusions at the beginning so that you can have a framework for comprehending the more detailed description of the evidence that will follow. My premise is that your knowing *why* the various levels of information are presented will help you make more sense of the flow of the data and evidence.

As *The Seeker and The Teacher of Light* makes clear, during my past 20 years as a Seeker, synchronicity has led me to see the world in a different light from the one in which I grew up and was trained. I view synchronicity as a higher guidance pointing me to follow a certain path, one that will eventually lead me to a goal desired by my inner self. Because I have Free Will, I do not need to follow any set course.

However, I have found that following the path of my intuition/higher guidance leads to satisfying outcomes. In following the path to which synchronicity has led me, I have come to a different understanding of how everything is put together in terms of who we are, creation, the afterlife, channeling, out-of-body traveling, and answers to many key mysteries.

#2. To Provide Tools for Investigating the Hidden World

I am a scientist, but in this book I will provide information about the nature of creation and life from another viewpoint than that of standard science. Even though I have made many observations on a variety of subjects by means of intuition and other modalities of direct perception, as a scientist I know that it is always good to have evidence. Standard science has many extremely useful tools for investigating the physical world; however, most of these tools are unable to detect the *hidden* world that forms the basis of the physical world. We all have intuition and other senses that give us information. For example, dowsers know where water is located deep in the earth by using special tools, such as pendulums, to focus on hidden energies to make such sites visible.

And so this book's second purpose is to provide you with tools to investigate the hidden world and bring it into the light for yourself.

In addition, I have come to understand *how to measure the energies* involved in many of the mysteries addressed in this book. These measurements afford me a still greater understanding of who we are, as well as of the validity of the views that I hold and will be sharing with you.

On a cosmic level, having such tools will enable you to gather information about the nature of matter and creation. For example, you can get an alternative glimpse into what happens to foods as you cook and digest them. You also will begin to understand the purpose of breath, beyond bringing in oxygen and breathing out carbon dioxide.

As you learn about the mysteries of life and how to connect with the many subtle energies and their fields that surround us, you will gain an understanding of your true I AM essence.

This can take a practical turn, as well. For example, you will learn to know what foods and supplements are good for you on a day-to-day basis.

The measurements that I have learned, you will also learn. Not only that, but the measurements themselves will begin to help you understand who you are. If you are interested in philosophy and religion, you will begin to understand the effects of some of the words and symbols that you use on an everyday basis. If you are a scientist trying to understand and work with plasma, monatomic elements, or gas in the nano state, you will be introduced to a new tool to measure the fields that they produce.

You will learn such things as:

- How to make activated/structured water that would be better for you than non-activated/less structured water.
- That everything is "alive" and constructed in the same way, and that you can communicate and work with their energies.
- How to create subtle energies that are harmonizing/beneficial for you.
- How, using just intention, you can instruct subtle energies to move to different locations.
- How to detect and measure subtle energies.
- How to create subtle energy fields.
- How to measure your vibrational energy level, and how to increase it to the I AM level.
- How to bring yourself into harmony.
- How to determine whether something is in resonance with you (i.e., good for you). This "something" can be as varied as electromagnetic fields (EMFs), foods, supplements, minerals, gemstones, essential oils, and many other items.

The Intersection of Science, Subtle Energies, Consciousness/Intention, and Spirituality

In this section, we will see how science, subtle energies, consciousness/intention, and spirituality come together.

Let's examine the evidence from each area.

CHAPTER 1

Applying Scientific Hypotheses to Subtle Energy, Consciousness, and Intention

Science shows that we are much more than just matter. Our very own thoughts and intentions can affect matter and energy. We can communicate with matter and other life forms.

Science is based on experiments designed to prove hypotheses. A hypothesis is "a proposed explanation for a phenomenon. For a hypothesis to be scientific, the scientific method requires that one can test it. Scientists generally base scientific hypotheses on previous observations that cannot satisfactorily be explained with the available scientific theories." (Wikipedia)

If evidence is reproducible, this means that it follows scientific principles of proof, *even if the evidence may not be accepted by some people because it contradicts standard beliefs.*

In the scientific mainstream, there are certain things that are not readily known about, or even believed. However, this does not mean that they don't exist or that they don't work powerfully behind the scenes.

Here are some examples:

- Subtle energy and subtle matter
- The spiral nature of the universe
- Resonance

Subtle Energy and Subtle Matter

Subtle energy and subtle matter exist, even if not known to or accepted by the mindset that views everything as having to exist on a physical level in order to be taken seriously. There are now thousands of persons who tune into the energies that surround us and who can detect and measure those energies. They do this by coming into resonance with those energies, using tools such as pendulums to help them focus on the wavelength of those energies. Scientists have been using various experimental methods to detect and measure subtle matter.

The ability to come into resonance with subtle energies, discussed in *The Seeker and The Teacher of Light*, is inherent in us and can be taught. It is like learning to walk or to ride a bicycle. Those who try will learn the skills to tune into the energies surrounding us.

The Spiral Nature of the Universe

Spiraling energies form the basis of matter creation and decay. *The Seeker and The Teacher of Light* described the basis of matter creation as light spiraling centripetally inward to form matter, with the outward spiral resulting in the disintegration of matter. These inward and outward spiraling lights describe a formation known as the *torus*. The center of the torus is the location of the fulcrum from which these opposing spiraling lights reach balance.

All matter is of this construction. Thus, everything is of this pattern, and we are created in this pattern.

Resonance

Resonance takes place when the wavelength of the vibration of one thing makes something with a similar nature vibrate without even being touched.

Imagine a monochord instrument where a string is attached, such as a violin. When you pluck that single string, it will vibrate and sound as a certain pitch, or note, is produced. The note created in this way has a wavelength to it, which defines that note. The vibration of the string you have plucked will be in *resonance* with other strings that share a proportional string-length (for example, twice as long; half as long; twice as short; half as short). And then they will all vibrate in multiples and octaves, ad infinitum.

So whatever is in resonance with you (for example, a food that is good for you) vibrates *in accordance with you* versus interfering with you. I consider that the scientific concept of resonance is real, as evidenced by tuning forks vibrating at higher and lower octaves due to resonance, and many other examples found in physics.

Four Key Hypotheses

The above forms the basis of four key hypotheses to be discussed here, as well as the evidence that supports the hypotheses.

The First Hypothesis: *Our mind/consciousness/thoughts/intentions can affect matter and energies*

The Evidence:

What is the evidence for this hypothesis? I will summarize it here, and also describe it in greater length later in this book, along with the abundance of evidence produced by other leading scientists:

1. *Mind/consciousness/intention can change the structure of water.*

   The mere act of holding your hands around or over water and giving it love and gratitude will change the structure of the water. There is a field called BioGeometry that allows you to detect and generate energy qualities that have a harmonizing effect. Through resonance, you can measure the subtle energy known as BG3, the three basic qualities that are in resonance with the one harmonizing subtle energy quality and are used to detect it. (More about this in Chapter 9, "The 3-6-9 Energies of GANS, Ormus, and Their Fields.")

   It was found that when the water is charged with love and gratitude for 1 minute, the BG3 level of the water will move up from around 200 to 1,200. A BG3 level of 200 (tap water) has a lower level of structure, as compared to a level of 1,200, which has the quality and structure of excellent spring water. (A BG3 Ruler is used to carry out the measurement.) An increase in BG3 is a reproducible fact, measurable by anyone trained to measure it.

   Activated water with higher BG3 levels has also been tested in the lab of Masaru Emoto, author of *The Hidden Messages in*

Water, revealing the production of very geometric hexagonal crystals. A BG3 water-activation level of 1,200 will produce a more geometric hexagonal structure when frozen, as compared to a BG3 level of 200. This fact can also be ascertained by tasting the blessed water, as compared to the water that has not been blessed.

2. *Subtle energy fields can be moved just by thought, and the moved field will activate water.*

You are familiar with magnetic fields, as can be visualized by iron filings on a piece of paper with a bar magnet under the paper. In the experiments that will be described in Chapter 3, "Resonance, Subtle Energies, and Tools for Their Detection," fields containing BG3 and 3-6-9 subtle energies can be formed from a variety of ways (GANS, Ormus, BG3 tools, 3-6-9 symbols, magnets). You can create a field with certain kinds of tools (GANS, Ormus, and 3-6-9 symbols). These fields can be detected and measured by BG3 or 3-6-9 pendulums and are scientifically reproducible.

These fields will also activate and change the *structure* of water. You can, by thought or vocally, say, "Move to _____," and the field will move to a location distal to the location of the initial field (i.e., away from its center). As a result, the BG3 energy that was in the original location is no longer there and is now in the new location, and is measurable in the new location. Since the energy has BG3 in the new location, you can put water in that new location and watch the BG3 level increase in the water. This is reproducible, and you can study the kinetics of the activation of water in the new field.

3. *Situations that are causing interference with intention can be harmonized.*

BioGeometry tools that result in an increase in BG3 often can be nullified by a 3-6-9 torus-based symbol (see Chapter 3). What is observable is that both BG3 and 3-6-9 energy qualities can no longer be detected. Simply the thoughts or words of inviting in harmony results in harmonizing the interference so that both BG3 and 3-6-9 energies are now present. This phenomenon is also reproducible.

4. *Intention can affect random chance.*

The most studied phenomenon of how intention can affect chance was carried out at Princeton Engineering Anomalies Research *(PEAR)* by Robert Jahn, Brenda Dunne, and Roger Nelson. Roger Nelson formed the Global Consciousness Project, in which hundreds of random-number generators (computers) randomly generating 1s and 0s every second were placed around the world. The appearance of 1s and 0s are random, and roughly the same.

However, when millions of minds are focused on an event, the numbers no longer remain random. This occurred with the 9/11 event, and also when Princess Diana died. These events synchronized millions of minds and caused a significant change in the random numbers. This could not have happened by chance, by the order of one in a trillion.

5. *Spoons can be bent by intention alone.*

People who have bent spoons and twisted them into a pretzel shape know that intention will result in a spoon's softening to allow it to bend easily. Children are often the best students at bending spoons (perhaps because they don't have fixed ideas that it isn't possible). You can watch the YouTube videos of Jack Houck, who popularized spoon-bending parties. I personally have bent many spoons. It requires a state of mind that allows the spoons to soften.

This is experimental proof. Those who have seen spoons soften know the validity of what they experienced.

6. *Accurate remote viewing is possible, even likely.*

The huge amount of experimental data collected by Russell Targ and others proves that we have the capability to remote view (see what is located at distant locations), as evidenced in the documentary, *Third Eye Spies*. Targ presents irrefutable evidence that we have the ability to remote view. Skilled remote viewers can describe with excellent accuracy what is located at another location. Distance does not matter. Time also does not matter, as exemplified by how Targ and his colleagues predicted the future of commodities markets of items they were remote viewing, days into the future.

7. *Intention can influence physiology.*

Dean Radin and his colleagues at IONS (Institute of Noetic Science) have demonstrated in many varied experiments that intention affects outcomes. In his book *Supernormal*, Radin provides evidence that a person's intention influences the physiology (e.g., skin conductance) of a distant person. The intentions of trained meditators have a somewhat larger effect than that of normal meditators.

These are just some of the experiments proving that we do communicate and affect matter and energies.

The Second Hypothesis: *We can communicate with our thoughts to matter, subtle energies, and other life forms.*

The Evidence:

1. *We are in communication with matter and energies.*

The evidence connected with the first hypothesis makes this clear. When you ask a spoon to bend, ask a BG3 subtle energy field to move, activate water, or affect random-number generators, you are communicating with matter and energies.

2. *We can communicate with plants and other life forms.*

The scientific experiments of Cleve Backster prove the ability to communicate with plants and other life forms. Backster, a pioneer in the use of lie detectors and the author of *Primary Perception: Biocommunication with Plants, Living Foods, and Human Cells*, hooked up a lie detector to a plant. The detector showed significant activity when Backster threatened harm to the plant (e.g., to burn the leaf). Even distance did not matter. Backster could be miles away and the plant would respond. In later years, another scientist working at IBM, Marcel Vogel, demonstrated the same phenomenon. He found that vocally expelling air as the threat was made accentuated the effect.

3. *We can communicate with subtle energies.*

As also described in the First Hypothesis, we can communicate with subtle energies. We can create fields that contain BG3 or 3-6-9, and ask the BG3 and 3-6-9 to move from one location to another location. We can test for the BG3 subtle energy that has been moved, and show that energy can activate water that has been placed in the area where the energy has been moved.

In another experiment, it was shown that 3-6-9 symbols can interfere with BG3 tools, nullifying the detection of both energies. Yet with simply the intention to harmonize the BG3 and 3-6-9 subtle energies, the interference can be harmonized and the energies can again be detected.

In Chapter 10, we will see that, with intention, we can copy subtle energies and place that copy onto another object. This phenomenon can be proven by testing for the subtle energy on the other object. When BG3 or 3-6-9 is copied onto a new object, the new object can activate/structure water, which can be measured.

The Third Hypothesis: *Subtle energy and subtle matter exist.*

The Evidence:

1. *The entire field of radiesthesia and BioGeometry show that we are detecting subtle energy or subtle matter when we detect BG3 and 3-6-9 (another subtle energy quality discussed in* The Seeker and The Teacher of Light*).*

All matter shows subtle energy qualities when a neutral pendulum rotates, once its wavelength has been achieved, as the pendulum is lowered from the user's fingers. Thousands of people now measure and detect BG3 routinely. I personally use radiesthesia every time I go shopping to find food or supplements that are in resonance with me. I do not depend upon a label to tell me if it is "organic," because I find that these labels do not give me the full answer. Instead, I depend on whether the food or supplement is in resonance with me, as shown by the clockwise rotation of the neutral pendulum.

"Radiesthesia," as defined by Dr. Ibrahim Karim, the founder of BioGeometry, is "the science of using the vibrational fields of the human body to access information about other objects of animate or inanimate nature by establishing resonance with their energy fields, using specially calibrated instruments and a scale of qualitative measurement to decode this information."

2. *Energies are seen between the hands when the hands are raised and face each other.*

   In a darkened room, you can see the energy beams cross the room from your fingers or from the palm of your hand.

3. *There is a body of scientific research into subtle energies.*

   Dr. Claude Swanson's book, *Life Force: The Scientific Basis,* Volume II, goes into great detail on experiments involving subtle energies. It recounts the names that different groups and investigators have given for the same or related energies—for example, *qi/chi, torsion energy, orgone, odic energy, prana, subtle energy,* and many others.

   Dr. Nikolai Kozyrev (1908–1983), a Soviet astrophysicist and astronomer, meticulously carried out a variety of experiments showing small weight or temperature changes due to torsion-energy changes. He observed left- or right-spiraling torsion energy due to entropy (decay), such as acetone evaporating, or due to negentropy (growth, life), in such areas as plants growing. Dr. Claude Swanson, quantum physicist and author, talks about these and other experiments under the general term of "subtle energy." Clairvoyants often see the subtle energies.

   Baron Dr. Carl (Karl) Ludwig von Reichenbach, a notable chemist, geologist, metallurgist, naturalist, industrialist, philosopher, and a member of the prestigious Prussian Academy of Sciences (1788–1869), described subtle energy as "positive and negative odic energies" (the term "subtle energy" had not been coined yet).

   Wilhelm Reich (1897–1957), the Austrian doctor of medicine and psychoanalyst who moved to the United Sates to escape the Nazis, discovered an energy that he called "orgone." He created "orgone accumulators" and "cloud busters" using these energies to cause rain. I highly recommend Dr. Swanson's book if you are interested in learning, from a scientific basis, about the various subtle-energy research carried out by a comprehensive list of experimenters.

There is the work of Dr. Klaus Volkamer, author of *Discovery of Subtle Matter: A Short Introduction,* who carried out 30 years of research measuring weight and temperature changes resulting from subtle matter. His research provides proof of the phenomenon of subtle matter. Some of his experiments will be described later on in this book.

The Fourth Hypothesis: *Spiraling energies form the basis of creation or decay of matter.*

The Evidence:

1. *Evidence from Walter Russell's work (described in The Seeker and The Teacher of Light).*

   From Russell, we learned that clockwise rotations are associated with matter creation (negentropy), while counterclockwise rotations are associated with breakdown of matter (entropy). The rotation of a pendulum is a reflection of the direction of spin for the spiraling energies of light discussed by Walter Russell.

2. *Evidence from the disciplines of radiesthesia and in BioGeometry.*

   From radiesthesia and BioGeometry, we observe the direction of rotation of the pendulum. We learn that things that are not good for us (e.g., foods, minerals, supplements) cause a counterclockwise rotation, whereas materials that are beneficial for us result in a clockwise rotation. This is consistent with the spiraling clockwise direction for creation, and the counterclockwise direction for the breakdown of matter.

3. *Things that are in decay (entropy) show a counterclockwise spin of the neutral pendulum, and things that are growing or organizing (negentropy) show a clockwise rotation.*

   This is true according to my own observation. My testing involved the same list of experiments carried out by Dr. Kozyrev for torsion energy, and my test results agree with his findings. This is also consistent with a spiraling clockwise direction for creation, and a counterclockwise direction for breakdown of matter.

4. *All of nature is consistent with spiraling energies manifested in physical form.*

The spiral resembles that of the Fibonacci spiral. This is seen in the spirals of pine cones, the shape of sea shells, the branching nature of trees, and even the shape of a human being (all the segments of the body are in the Fibonacci spiral shape). In mathematics, each number of the **Fibonacci sequence** is formed as the sum of the two preceding ones, starting from 0 and 1. Thus, part of the sequence is: 0, 1, 1, 2, 3, 5, 8, 13, 21, 34, 55, 89, 144, . . .

One can form a tiling with squares whose side lengths are successive Fibonacci numbers: 1, 1, 2, 3, 5, 8, 13 and 21. The Fibonacci spiral is formed by drawing circular arcs connecting the opposite corners of squares in the Fibonacci tiling. The Fibonacci is an approximation of the golden spiral.

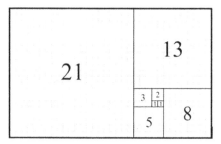 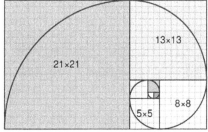

FIGURE 1. The above shows a tiling with squares whose side lengths are successive Fibonacci numbers: 1, 1, 2, 3, 5, 8, 13 and 21. The Fibonacci spiral is an approximation of the golden spiral created by drawing circular arcs connecting the opposite corners of squares in the Fibonacci tiling.

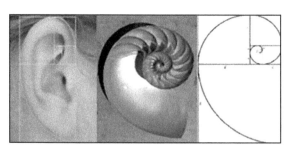 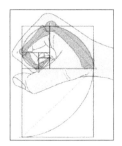

FIGURE 2. The above figures illustrate that the spiral is inherent in us and in nature.

The science in this book will show you, in reproducible experiments, that we are much more than just physical matter. We are all creators who can affect matter and energies with intention and thought.

Summary

Science is based on making observations, creating hypotheses to explain the observations, and collecting data to test the hypothesis. In this chapter, we have presented the hypotheses that:

- Our mind/consciousness/thoughts/intentions can affect matter and energies.
- We can communicate with our thoughts to matter, to subtle energies, and to other life forms.
- Subtle energy and subtle matter exist.
- Spiraling energies form the basis of creation or decay of matter.
- Evidence indicates that these hypotheses are correct. The rest of this book will go into more detail on the evidence. The evidence also provides data as to the nature of who we are—that our essence is I AM.

I AM, The Basis of a New Spirituality

This book pulls together the scientific study of certain subtle energies, the understanding of creation principles, and the understanding of our I AM nature. Chapter 2 presents a brief overview, to be covered in more detail in subsequent chapters. This will enable us to realize that I AM is the basis of a new spirituality.

The Spiraling Energies of Light—The Basis of All Creation

The new science states that:

The creation of matter is based on the compression of the clockwise spiraling energies of light to form matter.

These clockwise spiraling energies show up as a shape called "the torus structure." A *torus*, which has the appearance of a donut, is the shape of spiraling light (energy) compressing into a center. These are spirals that compress down to a denser portion in the center, which is the process of creating the densities that we experience as matter.

FIGURE 3. Model of a torus.

There are two aspects of the lights that spiral toward each other, forming a *double torus*. The interactions and compression of the spiraling torus lights result in the creation of the matter that we see.

Everything we see is of this composition. The positions and pressures in these spiraling light are what create the different chemical elements of which we are composed. Each element is similarly composed of these spiraling lights.

The torus is the main structure for creation. The spiraling energies of the torus are seen in everything that is matter. We even occupy a body that is composed of these spiraling energies.

Everything is a cycle. The first half of the cycle is *creation*, where the compression of the spiraling light takes place in a *clockwise* spiral. The second half of the cycle is the *decompression* (unwinding) of the light, in a *counterclockwise* spiral, which results in the radiative breakdown of matter. ("Radiative" means having rays or parts proceeding from a center.) Because this is a cycle, the light that has been unwound will recycle to again form matter by the compression of spiraling light.

Together, the two halves of the cycle form the life in which we exist. These two halves are two polarities of a single process. In a similar way, we might talk (as Walter Russell did) about the two spirals of light being "male" and "female" (another set of polarities that seek balance), of red and blue light, respectively.

This is the nature of the matter of our universe and everything in it. Every atom is made of spiraling light. One could say that the spiraling energies move to form a "black hole." Scientists such as Nassim Haramein say that every atom is centered by a black hole, which may suggest an analogy of light spiraling toward the center to form matter, with light spiraling toward the black hole at the center of the galaxy.

Walter Russell (1871–1963) was a polymath who excelled as a painter, sculptor, author, philosopher, musician, champion figure skater, and scientist. According to Russell's terminology, the first half of the creation/decay cycle is generative and electric, compressing light into matter; the second half of the cycle is the breakdown or unwinding of matter and is radiative and magnetic. All matter, Russell says, is based on the motion of light and the waves they create.

We are thus made of the compressed motion of light forming waves and having existence as matter, based on an equilibrium of opposite motions of light as they meet, establish the illusion of stability, pass through each other, and decay.

The Rainbow Body

The Rainbow Body is an interesting bit of esoteric information that provides a glimpse as to the light nature of who we are. This information is not meant to be proof of any hypothesis; it is just intriguing, since it reflects Walter Russell's concept that matter is created from light. Much of the Rainbow Body information is found in David Wilcock's book, *The Source Field Investigations.*

In Tibetan Buddhism, there are many reports of a Rainbow Body that is formed as part of the "death" process. Over many centuries, in China and Tibet, records indicate that there are over 100,000 persons who have achieved the state leading to the Rainbow Body. The Dzogchen monastery had 60,000 practitioners of Dzogchen reach the Rainbow Body state since its establishment in the 17th century.

IONS (the Institute of Noetic Sciences), working with David Steindl-Rast, a Benedictine monk, hired Father Francis Tiso, a Catholic priest who frequently visited Tibet, to investigate and document the Rainbow Body phenomenon. He interviewed many witnesses who recorded the death of Khenp A-chos, a monk from Kham, Tibet, who died in 1998. Witnesses stated that a rainbow appeared over the monk before he died, and that dozens of rainbows appeared in the sky after he died. He was not sick, and nothing appeared to be wrong with him as he chanted a mantra. After his last breath, his skin became pinkish to brilliant white. All witnesses said the body started to shine. They put the body in a yellow robe and watched as his body began to shrink. They heard music from the sky and smelled perfume. After seven days, his robe was removed and there was no body. Various persons stated that the monk appeared in their visions and dreams.

The practitioners, who are of a loving nature, practice *Dzogchen.* Dzogchen is the natural, primordial state of our being. Its teachings and meditations are aimed at achieving an all-encompassing primordial, timeless clarity that has no form, yet is the nature of the Universe and is capable of perceiving, experiencing, reflecting, or expressing all form. It is like a mirror that reflects with complete openness, but is not affected by the reflections. (Visualize your mind like a mirror all around you and understand that it is *all* you.)

After the practice of Dzogchen, there is the further practice of *Trek-cho*, whereby the mirror-like clarity of the Universe creates a knowledge called *rigpa* (becoming aware of God consciousness). Trekcho is an advanced practice that is the fastest way to achieve rigpa.

I AM and Balance

God is in everything and *is* everything. *God is at the center, controlling and balancing the spiraling lights.*

The oft-quoted phrase, "as above, so below," means that God is the mind, the consciousness in everything. Thus, *your* essence is God. And as God is the fundamental I AM, your essence is the I AM. (The meaning of the primary Hebrew name of God—Jehovah, or Yahweh; literally, the letters Yud Hay Vov Hay—is "I AM THAT I AM.")

I AM is at the fulcrum of the lights, balancing everything. Balance is found at the center. When everything is in balance, there is the kind of calmness and stillness that is achieved in deep meditation. When we deviate from the center, there is less balance, and therefore greater unhappiness/disharmony. When there is anger or fear, there is greater disharmony/imbalance and therefore less happiness.

Everything seeks the state of balance. Wars, disagreements, conflictual relationships—anything that does not take balance into account results in unhappiness/disharmony.

Once we begin to open to this reality, why would we *not* seek balance?

We Are Creators

In creating humanity, God has given us free will. This means that we have the ability to create the situations that cause us happiness or *un*happiness.

The primary way we create is through our thoughts. Our thoughts create everything. And although there is a collective creation and a collective consciousness, we are each Creators, and so we create our own reality.

When we come to realize that our thoughts create reality, we also realize that we have a choice. We can create a Golden Age, based on our realization of our connection with God and one another, or we can create in ways that do not take our Oneness into account but instead focus on separation and having power over others. We can literally create our own hell, if we choose to do so—it is not an outside source doing it for us. We all have Free Will.

The analogy is often made that life is like a school where we learn from experiences. From a spiritual perspective, we are presented with lessons of love, harmony, and balance. This is reflected in the Ho-opo-nopono teachings of the Hawaiian shamans: "Forgive me, I am sorry, thank you, I love you."

When you say these four simple statements, it means that you have taken on the responsibility for others' suffering as well as your own. It means that you are moving the situation being addressed by these statements toward the center, toward balance, and toward the realization of Oneness. Life often throws us situations that we have to deal with (e.g., abuse, injustice, wars, relationship incompatibilities, etc.). We grow not only by addressing those problems, but also by *how* we address them.

Sometimes, through karma, we are given experiences from which we can learn. Each lesson may offer us different ways of addressing the experience. Sometimes the lesson may be how to extricate ourselves and not put ourselves in harm's way, while at other times, the lesson may be to show compassion and love, and thus bring in harmony.

The new spirituality is knowing who you are—the I AM. The scientific data is now irrefutable that we are creators. The experiments I have carried out—some of which you can carry out for yourself later in this book—add further evidence as to the nature of who we are.

So the first step is to know this.

The next step is to boldly *state* who we really are. We each are I AM. We each are powerful divine beings. We are each creators.

When we realize that we are each I AM, we thereby move ourselves into the state of harmony and stillness, where all things are possible.

Secrets of the Ancient Symbols

Creation has been described by means of *key symbols and words* in the various main religions. Throughout the early history of religions and the world, the vocabulary, the understanding, and the science were not sufficiently advanced to fully explain the nature of creation. But now, the advances in science and history have progressed to the stage where anyone can access the information of the past, as well as the information of this age, since the Internet has put almost unlimited information at our fingertips. The general level of consciousness also has advanced sufficiently that key thoughts in this area are available to everyone.

At this time, through synchronicity or divine guidance, a new set of understandings has been revealed to me about the secrets of the ancient symbols (which are part of the key religions) and how to measure the power of these symbols. I believe that—given my background as a scientist and my dedicated interest in spirituality—I was led to study and understand the secrets of radiesthesia and BioGeometry. I have the privilege of introducing these technologies to you, so that you too can measure and detect the energies that surround us. In doing so, you may be transformed by coming to a deeper understanding of their essence, the nature of creation, and how you are a part of that creation.

The Torus and the Key Symbols of Creation

The key symbols of creation all point to the *torus* being the basis of creation. The same secret that's behind the torus structure is also the secret behind Aum, the bagua, and many other symbols:

Aum Bagua Yin yang Cross section – torus 369 as numbers Reiki power Vortex Math (torus)
See 369 within torus and as units symbol and as units

FIGURE 4. The key symbols of creation.

- *Aum*—a key symbol, chant, and ending word in prayers and meditations, in many religions (e.g., Christianity, Islam, Judaism, Hinduism). The Sanskrit aum is derived from 3-6-9.
- The *bagua* symbol and the related *I Ching* in Asian cultures.
- The *yin yang* is another symbol that can be used to bring in harmony with our yin and yang natures, and is also derived from the torus.
- Torus and 3-6-9 (the mystery of 3-6-9)—in the current era, Nikola Tesla stated, "If you only knew the magnificence of the 3, 6 and 9, then you would have a key to the universe."
- The Reiki power symbol, which has both BG3 and 3-6-9, and shows the direction of rotation of creation within the torus.
- There is a math that forms the torus structure, known as Marko Rodin's Vortex Math—later called "the Symbol of Enlightenment"

To many, these symbols are just words, chants, or drawings. However, they are much more. They have power. *The power can be detected in the subtle energies they possess.*

The *detection* of their energies can be measured by anyone with a BG3 and a 3-6-9 pendulum. The *magnitude* of the energies can be measured by using the BG3 ruler. This will be described in a later chapter.

The Benefit of Detecting and Measuring These Energies

What is the *benefit* of detecting and measuring these energies?

You can use these energies to structure water, as evidenced by the significant increase in the level of BG3 and 3-6-9 in water. Masaru Emoto, in his beautifully illustrated book, *The Hidden Messages in Water*, showed that water structured itself to produce beautiful hexagonal crystals when frozen, if the water was imprinted with positive messages such as Love. Dr. Karim, working with Emoto's lab, showed that the same beautiful hexagonal crystals with water were harmonized with BG3. The studies show that BG3 can be significantly increased with the energies of the symbols. These energies can be magnified significantly for harmonizing and for healing purposes. The symbols can be used to harmonize EMF (electromagnetic fields).

Science attempts to validate hypotheses by means of experimental data. The data shows that structures derived from various elements of the torus have measurable energies, and that many people can make these measurements. The measurements are reproducible, and thus fulfill one of the basic tenets of science. (These measurements are described in further detail later in this book.)

Also, when you bring yourself to your higher I AM vibrational level and verify it with a pendulum—that is, if it rotates clockwise rapidly—you have experiential evidence as to your true nature. You innately know and remember who you are: I AM. This point will be described further in the next chapter.

Our Abilities

We are each I AM and creators. Later chapters will discuss experiments where we can activate water and demonstrate the quantitative level of activation. Other experiments will show how we can harmonize interfering subtle energies just by inviting in harmony. Still other experiments will show that simply by thinking and inviting measurable subtle energies to move, those energies will move. We create with our thoughts. We *are* our thoughts.

In the meantime, before we describe the quantitative experiments with water, you can carry out the qualitative exercise below.

Exercise:
1. Taste some tap water in a glass. Then put your hands around the glass of water and bless it with love and gratitude for 1 minute.
2. Then slowly sip some of the blessed water.
3. How does it taste? Is it better than the original water? Does it taste more like good-quality spring water?

Our abilities to activate water, move fields anywhere in the world, and use symbols to activate water and more can be scientifically proven. The changes that result are reproducible, and anyone can do it.

Proof of Consciousness and Its Relationship to Matter

Scientists such as Dean Radin, Robert Jahn, Roger Nelson, Brenda Dunne, and others have proven that mind/thought/consciousness can influence matter. They proved this in their random-number-generator experiments causing deviations from chance by intention.

Distant healing through intention is real. Self-healing through intention is real. Remote viewing (the ability to pick up what is happening beyond the parameters of our physical sight) is real, as Russell Targ and Stephan Schwartz have proven. (Though something of a well-kept secret, our governments and other governments are using those abilities for gathering intelligence data.)

The list of studies in these areas is extensive. Much of this information has not fully entered into the consciousness of the population. But the data is there and, with the new information age, it will become known, especially to the younger generation versed in obtaining information through the Internet.

Each Part of God Has Consciousness

We are each of a divine origin. We are each one with God. It is our *essence* that occupies our bodies. Each cell, each organ, each bacterium, each atom within our body has its own essence, which is a part of God. Each has its own consciousness. Its overall awareness may not be as high as our awareness, but there is a level of consciousness.

Therefore, when you invite cells that are causing you a problem (disease, pain) to come into harmony with their disharmony—for example, as shown in Joachim Wippich's affirmation (in *The Seeker and The Teacher of Light*), "I invite the cells in my _____ [e.g., arm] to come into harmony with its disharmony"—these cells realize that they are in a state of disharmony, and they revert back into the state of harmony; and you are healed.

When you are in your I AM state, the cells will hear you. But if you are not present in your I AM state, then when you speak and think, the clarity of your message may be diminished.

We Have Always Been Creators—Now We Can Create Consciously

We have always been creators. That is how God made us. And we each have free will. In our evolution from animal to aware human, our creative abilities led us to want to dominate. We have often been driven by greed and the desire to take what we can, regardless of its impact on others and the environment. Then, when disaster falls because we have upset the natural balance and harmony, we often state that it is the fault of another group of humans. Or we accuse God for our miseries.

The truth is that we are *all* God. We were made by God to be able to create, to experience, to learn, and to evolve.

In the new spirituality, we need to recognize that we are creators and that we need to create in a way that brings in harmony and balance. Imbalances can and will occur. Disharmony will arise. Our physical selves are made of polar aspects, male and female, and we are each individuals. Yet as creators, our responsibility is to create in harmony and love.

"By Invitation Only"

As you go about your daily tasks, think of yourself as a creator. As a creator, your each and every thought, word, and action will have a consequence. For example, as soon as you try to impose your will on someone else and the other person does not *want* that will to be imposed, you have created an imbalance.

Thus, the secret is to *invite* people to do something. If it is their will do to it, you are not imposing your will. If there is still disagreement, then try to achieve a balance so both parties have enough of what they want so they are not either totally dissatisfied or totally satisfied—so that the compromises allow *co-existence with the disharmony*. Then, a workable balance will have been achieved.

Creating in Harmony and Love

We were placed on this school called Earth for a purpose. Earth is a dense reality, where it is difficult to create. This is largely a good thing, since many people have not learned to create in harmony, love, and balance. Thus, the creations of such people will cause problems for others, since it will infringe on others' free will. Force is then used, through various means, to impose the will of a person or group of persons on others. Indeed, our societies, governments, religions, educational systems, and so on have built-in mechanisms for the control of others.

Learning to create in love and harmony is one of the key lessons in life. This way of creating involves making decisions or requests with thoughts of love, harmony, and the principle of no-judgment present. This way of creating enables us to learn that there are many ways to perceive anything. If everyone would approach life with the intention of making their creations more harmonious, then wars, lawsuits, etc. would be greatly reduced.

Summary

In this chapter, we see that the creation of all matter is based on light spiraling toward a center. It is interesting to note that Tibetan Buddhist monks who have been sufficiently advanced in their practices have been reported to revert back to light at death (the "Rainbow Body" observations). As we evolve in spirituality and in being our I AM essence, we learn the centering required to create balance and harmony. We learn that we are Creators, and that we create with our thoughts.

The ancient symbols associated with creation—such as aum, yin yang, and bagua—possess subtle energies, which are harmonizing. Newer versions of the same energies are associated with the torus structure; the energies of 3-6-9 are derived from the torus. These symbols also have a balancing/harmonizing effect on us.

As we learn to detect and measure these energies, we learn more about our I AM nature. We learn that consciousness affects matter. We learn that everything has consciousness to varying degrees; we have the ability of awareness with our consciousness. We become aware that we are Creators, and that now we need to learn to create in Love and Harmony.

PART II

Basis of Measurements of Subtle Energies

In this section, you will learn about the principles of detecting and measuring subtle energies using tools that come into resonance with the energies. Practical exercises are presented for learning the techniques used for subtle energy detection. This section also describes the concept of coming into resonance with your own personal subtle energies, as well as the energies of BG3 and 3-6-9.

CHAPTER 3

Resonance, Subtle Energies, and Tools for Their Detection

Before going further, let's take a look at the world of vibrational energy and the fields and waves it produces. It is a world that is typically ignored by mainstream science. However, in the future, mainstream science must embrace it, since it provides the basis of understanding energies and health/healing. By reading this book and knowing how to take measurements, you can learn the power of the field and know for yourself the validity of what you observe.

The electromagnetic spectrum is one part of that world. Modern science works within that spectrum, and has many tools that can detect and measure energies in the vibrational energy world. Below is a common drawing describing that world.

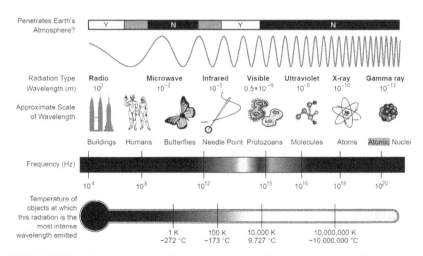

FIGURE 5. The Electromagnetic Spectrum as viewed by science today. Humans see only in a narrow band of the visible spectrum.

It should be noted that much of that world was only discovered starting in the 1800s—that is, not all that long ago. And yet our modern world would be lost without those discoveries—no television, no radio, no cell phones, no Internet. A person living in the early 1800s would think that what we do with our technology is magic!

The New Vibrational Energies

With the onset of radiesthesia, a whole new set of vibrational energies was discovered. It is similar to Madam Curie discovering radium: radium glowed and had various interesting properties, but until its discovery we did not yet have the *concept* of radioactivity, or an awareness of its dangers and benefits.

Similarly, the scientists studying the new vibrational energies did not fully understand them, or their dangers and benefits. In fact, one of the key scientists in this field, Leon de Chaumery, figured out how to generate large levels of a subtle energy (now called "vertical negative green"), and was killed by working with that energy. His friend and fellow researcher, Antoine De Belizal, found his body mummified by the effects of the energy.

The early scientists in this area had to devise ways to classify the energies, and determine which of the energies classified were present in various parts of the world that they were studying. The scientists knew they could use a *neutral pendulum* to detect the subtle energies associated with the various colors of the spectrum.

In the following sections, you will be taught how to make a neutral pendulum of your own, and how to use it to measure and detect subtle energies. Using this neutral pendulum, you will learn many new applications that are practical for everyday living, such as what foods and supplements are good for you and how you can determine that information.

Seeing into the Invisible World

To look at the invisible world from which the physical world is made, you need to "see" it. This chapter suggests tools that can be used to see the effects of the invisible world. The principle behind the tool is resonance. But the real tool is *you* and your ability to come into resonance with subtle energies, using appropriate tools.

The reality is that you are exposed to all sorts of energies that you can sense but cannot see with your eyes. In fact, there is such a wide range of energies whose effects show up in our world that your conscious mind most likely dismisses those energies. Therefore, you do not know they are there.

But what if you *could*?

The key is to have a tool that can focus on narrow bands of energy so that you can learn to "see" and measure them. As mentioned, the underlying principle behind the measuring tool is *resonance*. The *pendulum* is the main tool you will use for focusing your attention. The energies that the pendulum will detect are the subtle energies produced by virtually everything.

Once you have the opportunity to understand this, then we will delve further into the world of subtle energies and look at a variety of applications; this will give you a better understanding of the nature of matter and creation. Some of the concepts described in this section were first introduced in *The Seeker and The Teacher of Light*. In this book, these concepts will be gone into further (including applications) and will introduce you to a different way of viewing the world.

This chapter will provide you with the following foundation for developing your understanding and experiential relationship with resonance and subtle energies:

- A background for understanding the principles of detecting and measuring subtle energies, using the string length of a pendulum as a measure of wavelength to achieve resonance with the energy.
- Tools and experience in measuring and detecting subtle energies, including the neutral pendulum, BG3 pendulum, and 3-6-9 pendulum.
- Instruction on coming into harmony and the I AM level for optimal vibrational level before using a pendulum.

- Instruction on making your own neutral pendulum.
- Instruction on how to measure your own personal wavelength.
- Subtle energy qualities measured in this chapter: personal wavelength, BG3, 3-6-9.
- Practical experience using the personal wavelength, so you can know which foods, drugs, and supplements are in resonance with you (good for you), not in resonance with you (not good for you), or neither good nor bad for you. Also, testing EMF to understand when it is not in resonance with you (detrimental).

Understanding Resonance

Resonance occurs when there are vibrations, and those vibrations induce other vibrations at a distant location.

The classical example is the tuning fork. When a tuning fork is struck, its vibrations are picked up by other tuning forks of the same or different octaves, and those other tuning forks also will vibrate.

The same phenomenon is true for a string instrument. If you pluck the single string of a monochord instrument, the string will vibrate. (The vibrational frequency is dependent on length of the instrument's string/wire.) Then, other monochord instruments—either of the same octave or different octaves—will also vibrate. This phenomenon is illustrated in the figure below:

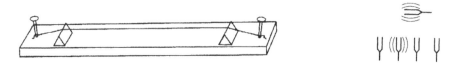

FIGURE 6. Monochord and Tuning Forks: Through resonance, vibrational energy information is communicated, so that other strings or tuning forks at different octaves also vibrate.

Understanding the Pendulum and the Wavelength of Vibration

A pendulum is similar to a monochord instrument, in that the length of the string produces the vibrational quality of the note when the string is plucked. The "string length" is defined as the *wavelength*. With a pendulum, the wavelength is the length of the string between (1) the index finger and thumb, and (2) the weight on the pendulum. If the wavelength is at an octave that's related to the object that is vibrating, then the pendulum will rotate.

You can find the correct wavelength for vibration to occur simply by initiating the swing of a pendulum and gradually lowering the weight that's attached to the string. When the string length is at an octave with the vibration of the object you are measuring, the pendulum will rotate clockwise. This concept is shown in the figure below.

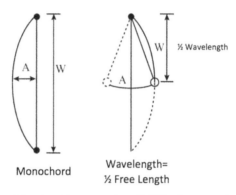

FIGURE 7. The string length or wavelength (W) of a monochord instrument is the same concept as the string length or wavelength (W) of a pendulum. Through resonance, the string length of the pendulum can detect vibrations of different octaves of the object being detected or measured. The strength of the vibration is the amplitude (A) and can be picked up by the diameter of rotation of the pendulum.

In this book, you will learn to detect and measure vibrations of subtle energies as they come into resonance with you, based on the string length of the pendulum. Your mind/body complex picks up the resonance and sends the signals to your fingers, resulting in rotation of the pendulum. This occurs at a subconscious level.

Array of Colors in Radiesthesia: The Basis of the Vibrational Energy Spectrum

If you bring a sphere outdoors, the spectrum of colors will be found around the sphere, based on the position of the sun, as shown in the figure below. What you are finding is the *resonance* of the colors around the sphere. If you draw a circle, then the resonance with the various colors is found arranged according to the directions of the compass. Thus, green is always at north, violet at east, red at west, and something we call "negative green" at south. There are also other colors associated with the chart, with whites, grays, blacks, and negative green. (Incidentally, negative green is not a color. It is called that because its location in the sphere or circle is opposite green.) Negative green has a penetrating power, since it traverses the entire sphere from the green position to the negative green position opposite green.

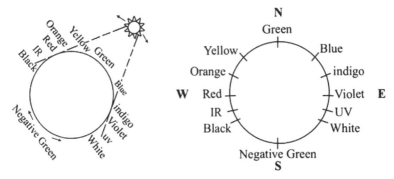

FIGURE 8. The diagram on the left illustrates the distribution of 12 colors around a sphere based on the position of the sun. This represents the classification that radiesthesia uses to describe resonance with subtle energies. The diagram on the right shows the same distribution, but based on the compass direction on a circle. Negative green is not a real color; it represents the vibrations of the penetrating energy *opposite* green.

It also turns out that each color has two directions of energy flow—a horizontal direction and a vertical direction. We won't go into detail about this phenomenon, except to say that very often the vertical colors are detrimental to life and the horizontal colors are positive to life. It was the *vertical* negative green that resulted in the death of de Chaumery, the French radiesthesia scientist, and left him mummified, as described earlier in this chapter.

In radiesthesia, each of the 12 bands of color can be divided into 12 sub-bands, resulting in an array of 144 bands to define the vibrational energy spectrum. One can then use the spectrum of colors to classify other energies or objects.

The Ability to Detect Subtle Energies

Matter, constructs (e.g., angles, shapes, symbols), colors, sounds, and many other things have subtle energies, which can be detected and measured with a pendulum. As you come to understand these energies and are able to detect and measure them, you receive tangible evidence of the nature of creation and who you really are. Having this direct experience will give you more valuable information than simply reading something and wondering what degree of truth it has.

So how does all this work?

It can be said that our bodies are created of light and energy, in a form called matter. Einstein, with his famous $E = mc^2$ equation, clearly stated that energy and matter are related (energy = mass times the square of the speed of light). Walter Russell, who received an inspired understanding of the nature of creation, stated that matter is the spiraling compression of light into matter.

Therefore, being *of* energy/light, we are *sensitive to* energy/light. The energies we are all very aware of are those that we can see or feel or hear—for example, the energies of heat/cold, sound, and light. In addition to these, there are the energies that plants use to grow, and there are chemical energies such as ATP (adenosine triphosphate), whose breakdown to ADP (adenosine diphosphate) produce chemical energy for our bodies.

What we are less aware of are the more subtle energies. Nevertheless, they are there and our bodies can pick them up.

For example, a woman being stared at from across the street by a man will invariably turn around to see who is staring at her. She is aware of the attention. In addition, there are dowsers who can find water using tools such as Y-shaped tree branches (which often are high in water content), as well as a pendulum or L-rod (an L-shaped metal rod in a sleeve, which allows the metal rod to spin).

How does this happen? For many people, all this seems like magic, or they may believe that it just cannot be done. But it can, and the explanation is quite simple:

We are made of energy, and therefore we can sense energy.

There is so much energy/matter around that we just need a tool to help us focus on a *particular* energy.

How we focus is through the laws of resonance. All matter is vibrating. Resonance allows us to pick up the vibration of whatever we want to pick up.

In the dowsing example, the dowser—the person looking to find water—is mostly *made up of water*, and so is naturally in resonance with water. (In *The Seeker and The Teacher of Light*, I described dowsing and how to make a pendulum to help focus the power of resonance.) The body of a dowser already knows the frequency of water, so for that person the pendulum will rotate clockwise in resonance with the water, whether the water is in a glass of water or deep inside the earth.

Traditional dowsing typically looks for "yes and no" answers. This is often called *mental dowsing*, and seeks answers to questions. When looking for water, the dowser will combine elements of mental dowsing and elements of dowsing using resonance. Resonance with water allows the dowser to inherently know the location of water. However, when the person is doing mental dowsing, the signal results in a "yes/no" answer.

In this book, we will only use the techniques of radiesthesia. Radiesthesia does not depend on asking questions and getting yes/no answers. Instead, we merely watch the pendulum once a to-and-fro motion has been initiated, and see if there is a clockwise rotation. If there is, it indicates the presence of resonance with the subtle energy. You can use a pendulum to see if there is resonance with almost anything (e.g., water, minerals, or yourself).

Exercise: Using a vial as a pendulum

To illustrate this, I made a pendulum out of a plastic vial with a cap, and attached a string to the cap (see Figure 9). If you have a plastic vial (or a container on which you can attach a string) as shown in the figure below, you can do the experiment too.

Example 1:

1. Fill the vial, converted into a pendulum, with water. The pendulum is now constructed to be in resonance with water.

2. When the pendulum is placed over a glass or vial containing water, it will go into resonance with the water in the glass or vial.

3. After you have initiated a to-fro-motion of the pendulum, the pendulum will rotate clockwise because of resonance with the glass or vial of water.

Example 2:

1. Place an aspirin in a vial converted into a pendulum.

2. When the pendulum is placed over aspirins on a table, it will go into resonance with the aspirin.

3. After you have initiated a to-and-fro motion of the pendulum, the pendulum will rotate clockwise because of resonance with the aspirin on the table.

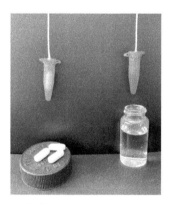

FIGURE 9. Pendulums made from vials with an aspirin and with water. These pendulums will rotate clockwise when they are held above either water or aspirin pills, demonstrating your ability to detect resonance using a focusing device (the pendulum).

Exercise: Practicing for Resonance with an Easy-to-Make Metal Pendulum

Since you may not have a vial in which you can attach a string to carry out the above examples with water or aspirin, you can carry out another exercise.

1. Tie a string to a metal nut. This creates a pendulum that would be able to come into resonance with another metal nut, washer, or bolt of the same type of metal.

2. Hold the string about 2 inches from the nut.

3. Practice a while to let your muscle memory get familiar with a to-and-fro motion, clockwise rotation, and counterclockwise rotation. Then create a to-and-fro motion with the pendulum over a nut, washer, or bolt made from the same metal as the nut on the string. As resonance is achieved, the pendulum will rotate in a clockwise direction.

4. If you now move the pendulum over another type of surface (e.g., wood table) and again start the to-and-fro movement, the pendulum will have just the to-and-fro motion, indicating that no resonance was achieved.

Below is a picture of a pendulum made from a nut and a washer, which can be placed on a table to check for clockwise rotation. If you have problems doing this exercise, the section after learning how to make a neutral pendulum ("Preparation Exercises Before Working with a Pendulum") describes how to overcome common blocks in this type of exercise and when using pendulums.

FIGURE 10. A pendulum made with a string and a nut. The nut is metal and the washer below the nut is metal. Because of resonance between the two similar metals, the pendulum swings clockwise once the to-and-fro motion has been initiated to overcome inertia.

The Long History and Modern Healing Uses of the Field of Radiesthesia

All this is not new. It is simply ancient knowledge that is being redis-covered. Practitioners of radiesthesia and BioGeometry work with the resonance/pendulum concept all the time. The ancient Egyptians used tools such as the Wadj, the Ankh, and Staffs with detectors at their ends for detecting energies. The Wadj is a dowsing tool, although many museums exhibiting the Wadj did not know its true purpose.

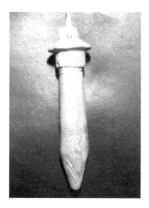

FIGURE 11. The Wadj dowsing tool from ancient Egypt.

Throughout ancient history, priests and shamans used dowsing tools to locate earth ley lines (earth grid lines linking sacred power spots) for locations on which to build their temples, churches, and cities. The French rediscovered this field in the mid-20th century and used it in medical practice to find the best remedies to treat patients. They named the field *radiesthesia*. It worked well for the patients, allowing better therapy by knowing what drugs or herbs would work and how much of the drug should be used. The physician would test to see if the drug was in resonance with the patient, as indicated by the clockwise rotation of the pendulum. The level of the drug could also be determined by reso-nance. But the medical community at that time did not understand the principles of the technology and banned it.

The Jesuits, too, made excellent use of the technology. Around the 1930s, the Jesuit priest Abbé Mermet wrote "Comment J'opere," in which he explained how he used radiesthesia to detect water and minerals onsite. He later wrote *Tele-Radiesthesia*, explaining how to find water and minerals at a distance. Most of the water sources around Paris and Vienna, for example, were discovered in the 18th century by famous practitioners of radiesthesia.

In the 1930–1940 timeframe, French scientists Leon de Chaumery, Antoine De Belizal, Louis Turenne, and radiesthesia scientist Scariatin (a Russian living in Egypt—pseudonym: Enel) advanced the science of radiesthesia by, for example, linking colors to give a nomenclature to classes of subtle energies, using a sphere. It was found that when the sphere was exposed to sunlight, different parts of the sphere would resonate with different colors (see Figure 8).

This is useful in any study of subtle energies. As an example, the essential oil lavender can also be described as having the subtle energy of infrared, as determined by its resonance with infrared on the circle or sphere.

During this same time period, the Germans further advanced the field by identifying the Hartman, Curry, and Benker earth-grid lines. Toxic energies can be produced by the crossings of these lines, especially if there is also a crossing of lines of underground water. Plants and trees do not grow well at these crossings, forming gnarled shapes or lack of growth. For humans living at such sites, cancers and other diseases can occur.

Through a series of synchronicities, a modern-day Egyptian scientist, Dr. Ibrahim Karim, came into possession of the valuable documents from the Maison de la Radiesthesia. With that store of knowledge and his own intuitive abilities, he discovered the secrets to the centering, harmonizing energies of BG3 and founded the science of BioGeometry. BioGeometry now has thousands of practitioners throughout the world, who, with their BioGeometry tools, are bringing greater harmony into the world. There are tools (BG16 pendulums) for detecting the harmonizing, centering energy quality of BG3, as described in *The Seeker and The Teacher of Light*.

There are a variety of tools to bring in harmony and centering to a person or to the environment, such as BioSignature pendants to resonate with the person; cubes to bring in BG3 to a house; strips and corner stands to harmonize earth-grid lines; as well as strips and other tools to harmonize sensitivity to EMF (phones, 4G and 5G signals). Other tools tap into the physical, vital, emotional, mental, and spiritual aspects of a person that may be causing problems, manifesting as physical problems. Most of these tools are used to detect the location of the problem, and/ or to correct those problems by bringing in harmonizing BG3 qualities.

Dr. Karim is well known for harmonizing the electromagnetic field (EMF) that had been causing illnesses among the population in two Swiss cities, Hemberg and Hirschberg (as discussed in *The Seeker and The Teacher of Light*).

It is from studying this rich background of information that I have been able to use BioGeometry tools, and to create tools such as neutral pendulums or pendulums to detect the 3-6-9 creation energy or torus energy. The tools detect resonance with subtle energies. The measurements they can make have given me greater insight into the workings of nature and our essence.

Radiesthesia and BioGeometry vs. Mental Dowsing

The science of BioGeometry and radiesthesia forms the basis of energy detection. It is different from standard mental dowsing, where questions are asked and "yes"/"no" answers are determined by the person's having trained the pendulum to move in a particular pattern for a "yes" answer or a "no" answer. Usually, a "yes" is trained to be a forward-and-back-ward to-and-fro motion. The answer that is usually trained to indicate "no" by the pendulum is typically a side-to-side motion. Both these directions are analogous to nodding the head up and down for a "yes" and turning the head right and left for a "no."

BioGeometry and radiesthesia work differently, using techniques to detect the presence of subtle energy qualities by resonance with the subtle energy. The person's use of the pendulum is not for the purpose of answering a "yes" or "no" question.

The mind of the dowser is blank—not occupied with getting an answer to a question, but simply in a state of awareness and observation.

Once the pendulum's to-and-fro motion has been initiated in order to begin the process (to overcome inertia), the dowser waits to observe whether a rotation of the pendulum occurs. If it does, this indicates the presence of the energy, as detected or measured by its resonance with the subtle energy.

Let us now look at the main tools we shall use and discuss in this book.

The Importance of Learning to Use a Neutral Pendulum

Readers of *The Seeker and The Teacher of Light* learned to use the neutral pendulum to measure vibrational level. If you were one such reader, you found that after the neutral pendulum has measured your personal wavelength, it will rotate clockwise rapidly as you raise your vibrational level by stating some affirmations for remembering/rethinking that your identity is I AM. (See the Affirmations by Joachim Wippich in the Bonus Chapters at the end of this book.) That alone is enough reason to learn to use the neutral pendulum.

But there is much more. We all shop for food and for various supplements. We all take various drugs for ailments. With the neutral pendulum at your personal wavelength, you can tell which foods are good for you (in resonance with you) in terms of benefits, which are bad for you, and which are neutral for you. The same is true for supplements, essential oils, minerals, and drugs.

Later in this chapter, you will learn how to find your personal wavelength with the neutral pendulum, and also learn more about its utility for many practical purposes. With the neutral pendulum, you can find the resonance of just about anything.

The chapter on dowsing in *The Seeker and The Teacher of Light* describes the neutral pendulum and how to make one.

But to make sure you have such a pendulum, I am repeating how to make it, below.

Making a Neutral Pendulum

One of the advantages of a neutral pendulum is that it will not resonate with other materials when used in radiesthesia testing—that is, it is *neutral*. I find radiesthesia techniques to have a high level of reliability, since they take the "thought" out of the method.

A neutral pendulum can be made of wood, acrylic, or another non-crystal material. Vesica Institute (*https://vesica.org*) sells neutral pendulums made of 1-inch acrylic balls with string. I have made neutral pendulums from self-hardening polymer clays, available at any craft store or online. Wooden pendulums are inexpensive and can be readily purchased online from a variety of stores.

However, the least expensive way (and possibly the quickest) is to make one yourself out of flour and salt, as described in the following recipe. This kind of pendulum works very well for the techniques described in this chapter. It is fairly light (a pendulum needs enough weight not to be totally weightless) and rotates well.

You can also go to any craft store and buy some self-drying clay or dough, or a type of dough or clay that dries by heating in an oven. They should all work for making a neutral pendulum. I used flour and salt, since it is typically available in most homes.

Recipe for making your own neutral pendulum from flour and salt:
The dough:

1. Mix ¼ cup salt and ½ cup flour with a little less than ¼ cup of water. (This is a consistency that works for me.)

2. Knead the dough.

3. Take enough dough to form a 1-inch ball (or any other shape you want).

4. Cut a string length of about 7 inches.

5. Tie a knot on the end of the string and insert it into the dough. (The knot helps prevent the string from pulling out of the pendulum after you form it.)

6. Then form the pendulum, keeping the string in the center of the ball of dough so that the pendulum is balanced.

7. Place the pendulum on some flour as a cushion to prevent it from deforming while drying.

The string:

For string (typically sold as twine), #12 size is a good thickness. Don't get tar-coated string; it is too stiff. Nylon string is also problematic: it tends to unwind, so you have to coat the ends with glue. Most personal wavelengths run around 1.5 inches (that is, the length of the string that corresponds to your resonant energy), so for practical purposes you don't need an excessive string length.

Let the pendulum dry:

1. Let the pendulum dry on its own for a few days.

2. At that point, you can (if you want to) tie a loop at the end of the string and place the loop on your pinky finger so that the excess string doesn't get in the way. But since you don't need an excessive length (as mentioned above), this is optional.

3. If you want to eliminate the rougher dough feel of the pendulum, you can give it a light coat of polyurethane (optional).

Below are a few pendulums I made using the above procedure.

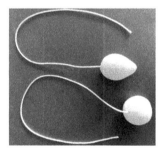

FIGURE 12. Pendulums made of salt, flour, water + string.

Preparation Exercises Before Working with a Pendulum

Before describing how to use a neutral pendulum for all sorts of applications, especially with your personal wavelength, I want to give you some simple exercises to make sure that energy is flowing smoothly in your body. For many of you, even the concept of energy flowing in your body is a different concept.

Exercise: Feeling Energy Sensations

1. Stare at your fingertips and mentally ask to feel the sensations in your fingertips. Do this for about a minute and experience whatever sensations you feel. If you are right-handed, be sure to also try this exercise with the left hand (I find that my left hand is more sensitive than my right hand).

2. Then place your fingers about 7 inches above a surface (table, chair) and see what sensations you feel in your fingertips. This gets you used to feeling energies.

3. Afterwards, you can put your left hand on your right shoulder and brush down with the left hand to the right hand. This reminds the body that you will be moving energy down the right hand to the right fingertips (for a right-handed person).

Exercise: Making Sure Your Polarity Is Not Switched

Occasionally a person's energy polarity is switched, so pendulums that should rotate clockwise will rotate counterclockwise. A simple physical movement usually corrects this phenomenon.

1. While standing, lift up your left knee high enough to tap the knee with your right hand.

2. Then lift up your right knee and tap with your left hand.

3. Repeat this motion 5 or 6 times.

It is like marching in place but touching the opposite knee each time, and at a speed similar to marching/walking. This usually corrects polarity switches.

Exercise: Training the Muscle Memory of Your Fingers with a Pendulum

Most people do not know the feel of a pendulum when it is rotating clockwise, counterclockwise, and making a to-and-fro motion.

1. Using the index finger and thumb, pinch the string of the pendulum about 2 inches from the dough ball.

2. Play with the pendulum and make it rotate clockwise and counterclockwise, let it move to-and-fro, away from you and towards you, and also move left and right.

3. Practice these series of motions with the pendulum until you are comfortable that your body remembers them.

Exercise: Affirmations to Achieve Harmony and Vibrational Levels Conducive to Working with a Pendulum Accurately

If you have done at least some radiesthesia or dowsing, the above exercises, starting with "Feeling Energy Sensations," are second nature and you do not need to do those exercises. For the novice, those exercises have value.

But very few practitioners of radiesthesia and dowsing fail to do what is described below. I generally do the exercise below before I do any radiesthesia work or any dowsing.

Many practitioners of radiesthesia and dowsing request that interfering thoughts be "cleared." However, bringing yourself into harmony is much better than trying to "clear." From my work with Joachim Wippich and from watching the enhanced skills that people develop when they use his techniques, I strongly recommend that you do the following procedures.

He first teaches us how to bring the two hemisphere of the brain to come into harmony with the heart. Then he teaches us to bring our vibrational state up to the I AM level. When this happens, energy is flowing in the body and you will find it easy to perform accurate radiesthesia and dowsing.

Affirmation to Harmonize/Balance Yourself (Right Brain / Left Brain / Heart Balance)

Hold your hands out to the side and then bring them together in front of your heart while saying each of the following lines:

Women:

"I AM inviting Every Thought within my Left Brain and my Right Brain to join together with my heart I AM"

Men:

"I AM inviting Every Thought within my Left Brain and my Right Brain to join together with my heart I AM"

This exercise can also be done every morning or to bring yourself back into balance any time you think you might be out of balance, not just before dowsing.

Bringing Yourself to the I AM Level ("I AM Thoughts I AM" and "I AM Everything I AM")

State the Affirmation below to bring yourself to the I AM level:

"I AM Rethinking Rethink I AM"

Pause

"I Am Rethinking Rethink I AM"

Pause

"I AM Rethinking Rethink I AM"

You pause between your repetitions so that you can absorb what you are saying.

If you are already a dowser, you can check your I AM level. Repeat the above affirmations until you reach 100% I AM. If you know how to check your personal wavelength, you can watch your pendulum swing faster and faster as you reach your I AM vibrational level. If not, don't worry; we will cover the subject again in the sections below.

Another statement I often use to reach the I AM vibrational level is:

"I AM Everything I AM"

As I repeat this statement, I reach the I AM vibrational level. It is the realization that we are One. It is similar to the analogy of a wave that thinks it is an individual and does not understand that it is part of the ocean. We are dense light beings in an ocean of light, imbued with the consciousness that has now grown to the state of awareness. We are the thought creation of God, and God is in everything. We are not separate from God or the One. We are thus not separate from our other selves who are reading this book. Thus, "I AM Everything I AM" has great meaning and brings us to the I AM vibrational level very quickly.

Exercise: Getting the Feel of a Pendulum Pulling in a Clockwise and Counterclockwise Manner by Testing Battery Polarity

1. Take out any battery (e.g., AA, C) and place it on its side on a table.

2. With your index finger and thumb, hold the string of the pendulum about 2 inches from the pendulum weight (the dough).

3. Start a forward-and-backward to-and-fro movement over the positive end of the battery.

4. Once the motion has been initiated, let the pendulum swing on its own (no attempt by you to make a to-and-fro motion). Just allow the pendulum to move freely on its own. You will notice that the pendulum, on its own, will rotate clockwise.

5. Bring the pendulum to the middle of the battery and start the to-and-fro motion. You will notice that it will just continue its to-and-fro motion.

6. Now bring the pendulum to the negative end of the battery and start the to-and-fro motion. If you allow the pendulum to move freely on its own, the pendulum will rotate counterclockwise on its own.

This exercise allows you to "feel" how a pendulum will move on its own without conscious control on your part. Your body "knows" the polarity of the battery by the gravitational pull of the positive pole making the pendulum rotate clockwise, or the magnetic pull of the negative pole making the pendulum rotate counterclockwise. That subconscious information is transmitted through your arm to your fingertips, and the rotation of the pendulum occurs naturally without conscious thought on your part.

Below is a picture of a battery standing up (rather than on its side, as described above). When you start the forward-and-backward motion of the pendulum, the pendulum will rotate clockwise over the positive pole and counterclockwise over the negative pole. Dr. Robert Gilbert at the

Vesica Institute has an excellent instructional YouTube video reviewing how to use a neutral pendulum: *https://youtu.be/rklULOn8p0w*.

FIGURE 13. For practice, when you hold the neutral pendulum above a battery and initiate the to-and-fro movement, the pendulum rotates clockwise over the positive pole and counterclockwise over the negative pole.

Applications Using a Neutral Pendulum

Now that you have developed a feel for the pendulum and are more familiar with the flowing energies in your body, let's apply the techniques to some measurements.

Detecting Resonance of Colors

Resonance with colors (which are simply wavelengths of light) can be easily detected. In this exercise, you will determine the length of the pendulum string corresponding to the wavelength of the color you are measuring. In essence, you are determining the string length that allows the pendulum to be in resonance with the color you are measuring.

Exercise:
See figure below for detecting the wavelength of red:

1. Assemble an object that has a color (e.g., colored paper) on a flat surface. It could be red, blue, or any other color.

2. Take your neutral pendulum, and hold the string between your index finger and thumb, with the weight next to the thumb/

index finger (i.e., put your thumb and index finger all the way down the string, just above the weight).

3. Start the forward-and-backward to-and-fro motion of the pendulum as you gradually lower the weight over the colored object. (You are actually sliding your fingers up on the string.).

4. Once the to-and-fro motion of the pendulum has been initiated, you no longer are consciously trying to make a to-and-fro motion. You are allowing the pendulum to move on its own.

5. The pendulum will change to a clockwise rotation at the wavelength of resonance with the color. That wavelength (string length) will now rotate clockwise over that color on any object. The length of string in which resonance occurs is approximately 1.5 inches. It usually takes me about 5 to 10 seconds to lower 1.5 inches on the pendulum string. Every different color will have its own wavelength, but they all are close to the 1.5 inches and are easily determined to be different from one another, since the string length that causes the pendulum to move clockwise for red is different from the string length that causes rotation for any other color. If you had tested for the string length for the color red, all other objects in your house that are red will also cause the pendulum to rotate clockwise. For any other color, the pendulum will just have the to-and-fro motion.

6. You can prove this for yourself by now bringing the neutral pendulum over another object with that same color. It will rotate clockwise. Then hold the pendulum over an object with a *different* color. The pendulum will just have the to-and-fro motion and will not move clockwise.

Feel free to repeat this experiment with other colors until you are comfortable with carrying out this exercise and you understand the concept of resonance with colors.

In the early part of this chapter, you learned that, in radiesthesia, resonance with the various colors around a sphere facing the sun or with a circle oriented according to the directions of the compass was used to define the vibrational spectrum of subtle energies. The example of finding resonance with the wavelength of the color red shows how the spectrum around the sphere or circle was determined.

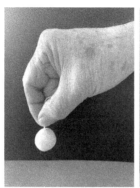

FIGURE 14. Wavelength of the color red. Using a neutral pendulum, gradually lower the pendulum by sliding up the string with the fingers over the red surface. Overcome inertia by creating a to-and-fro movement. When the string length reaches the wavelength of red, the pendulum will rotate clockwise. The pendulum will not rotate clockwise over other colors but will do so over any other object that is red.

The Neutral Pendulum and Your Personal Wavelength

A neutral pendulum is a pendulum made of a neutral material attached to a string. The neutral material can be, for example, a plastic ball (for example, acrylic); wood; or a weight made of flour, salt, and water (see recipe earlier in this chapter). All materials and electromagnetic energies—including the human body—are vibrating, and will thus resonate when they are at the wavelength that defines them.

Remember our earlier analogy of a monochord instrument (a single-string instrument that can be plucked to generate a note) or a tuning fork? When any note on the string or tuning fork is played, octaves above or below the note of the monochord or tuning fork will vibrate. This is because they are in *resonance* with the primary note.

The length of the string on the pendulum is analogous to the wavelength of the note being plucked on the monochord instrument. When you hold the string of a pendulum where it meets the pendulum body, and you gradually release the pendulum body and thereby extend the length of the string above the back of your hand, *the string length at which the pendulum starts rotating clockwise is called your "personal wavelength."* As you slowly lower the pendulum body on the string, you initiate a 45-degree swing to overcome inertia. The 45-degree swing changes into a clockwise rotation when the personal wavelength string-length is reached (usually around 1.5 inches).

What this means is that the pendulum is now in resonance with your body. Anything that is good for your body will resonate positively (that is, the rotation will go clockwise). If something is not healthy for you, the pendulum will move in a counterclockwise direction. If something is neutral for you (neither good nor harmful for you), the pendulum will just move back and forth.

Once I know my personal wavelength, I can do simple tests with a neutral pendulum to know what is good for me. I can go to the grocery store and know which vegetables, eggs, meats, and fruits are good for me. I can tell which vitamins or supplements will be effective for me. If someone says that a particular essential oil will be beneficial for me, I can tell if that is true or not. I also can tell if the microwave electromagnetic field of a smart meter or a cell phone will be detrimental to me.

The figure below shows examples of some neutral pendulums, one made of acrylic and the other of wood.

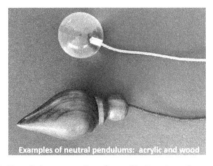

Examples of neutral pendulums: acrylic and wood

FIGURE 15. Pendulums made of acrylic (top) and wood (bottom)

Exercise: Finding Your Personal Wavelength—the String Length at Which Your Pendulum Rotates Clockwise

The following photos show how to lower/release a pendulum along the string to find the length at which the pendulum rotates clockwise. This is your personal wavelength.

  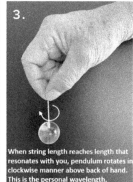

To start, hold string next to the pendulum above back of hand

Start slowly releasing pendulum on the string and moving pendulum back and forth above back of hand

When string length reaches length that resonates with you, pendulum rotates in clockwise manner above back of hand. This is the personal wavelength.

FIGURE 16. How to find a pendulum's clockwise-rotation length (personal wavelength)

1. Hold the string next to the pendulum above the back of the hand.

2. Slowly releasing the pendulum on the string, move it back and forth above the back of the hand.

3. When the string length reaches the resonant length of yourself, the pendulum will rotate in a clockwise manner above the back of the hand—your personal wavelength.

To summarize finding your personal wavelength:

- Hold pendulum string between the thumb and index finger at the base of the string and pendulum body.
- Hold pendulum above the back of the hand.
- Slowly release (lower) the pendulum over the back of the hand, while initiating a 45-degree swing to overcome inertia of a non-swinging pendulum.
- At about 1.5 inches, the pendulum starts rotating clockwise as it finds the length of string corresponding to your personal wavelength.

You can now check if you are in resonance with anything you want to test, by holding, touching, or pointing to the test object.

Resonance is indicated by a clockwise spin. A negative effect is indicated by a counterclockwise spin. A neutral effect is indicated by a back-and-forth motion.

Exercise: Practical Uses of Personal Wavelength

Foods

Foods that are in resonance with you are good for you. This statement means that:

- Foods that cause your neutral pendulum to rotate clockwise at the string length of your personal wavelength are positive for you.
- If a food causes the pendulum to go counterclockwise, the food is detrimental to you.

The directions below will show you how to determine what is healthy or not healthy for you.

1. Go to the refrigerator and pick out a sampling of foods.

2. Slowly release the string of your neutral pendulum over the back of your hand and initiate a to-and-fro movement (see instructions above on finding your personal wavelength).

3. Once the pendulum is moving to and fro, let it move on its own.

4. At around 5–10 seconds, the pendulum weight will have dropped around 1.5 inches. Usually, within the 5–10-second and the 1.25–2.0-inch range, the pendulum will start rotating clockwise on its own. The pendulum typically will also rotate a little beyond when the pendulum starts rotating, since there is a small range that is the resonant string length. Beyond that length, the pendulum stops rotating. The range in which your pendulum rotates is your personal wavelength. (The speed of rotation is also an indication of your vibrational level.) At twice the string length for personal wavelength, the pendulum will again rotate, since that is the next resonant octave. However, it is much easier to work with the first octave, since the string length is shorter.

5. Keep your thumb and index finger at the personal wavelength distance.

6. Put your non-dowsing/non-radiesthesia hand over a food. (Your left hand, if you are dowsing right-handed; your right hand, if you are dowsing with your left hand.) By doing so, through

your intention, you have linked yourself to the food in question. Although you could be touching the food, you do not have to. Just having your hand over the food connects you to the food.

7. Start the to-and-fro motion of the pendulum and then observe its motion.

   - The pendulum will pull into a clockwise rotation if the food is in resonance with you (i.e., good for you).
   - If the pendulum remains in a to-and-fro motion, the food is neither good nor bad for you. It is not benefiting you but it is also not harmful.
   - If the pendulum rotates counterclockwise, you should stay away from that food since it is detrimental for your body (i.e., not in resonance with you).

8. Repeat this procedure with other foods that you have brought out of the refrigerator. Typically, foods that have been grown in the presence of glyphosate (a detrimental herbicide sometimes used in non-organic farming) will cause the pendulum to rotate counterclockwise. If a food is contaminated, the pendulum will rotate counterclockwise. Do not eat those foods. If you have some older food in the refrigerator, test them to see if the pendulum goes clockwise or counterclockwise. This indicates whether you can still eat the food or should discard it (i.e., if the food is spoiled).

9. If you have any type of artificial sweetener, test it with the personal wavelength procedure, above. The pendulum will likely turn counterclockwise.

10. Use the above procedure whenever you go shopping. Perhaps counter to your expectations, non-organic foods are often okay for you (the pendulum will rotate clockwise) and some organic foods may sometimes cause the pendulum to go counterclockwise (i.e., not be good for you). What's behind this? You do not necessarily know the history of the food. It could have been grown in fields that previously used a

detrimental chemical, or there could be contamination in the water supply. Testing is the best way to determine if a food is good/healthy for you or not healthy for you.

Exercise: Supplements and Drugs

It is always good to know whether the supplements and drugs you are taking are in resonance with you. You can use your pendulum to find out.

1. Take out the drugs and supplements you normally use, and put them on a surface (e.g., a table).

2. Slowly release the string of your neutral pendulum over the back of your hand, and initiate a to-and-fro movement (see instructions above on finding your personal wavelength).

3. Once the pendulum is moving to-and-fro, let it move on its own.

4. After around 5-10 seconds, the pendulum weight will have dropped around 1.5 inches. Usually, within the 5–10 second and the 1.25–2.0-inch range, the pendulum will start rotating clockwise on its own. Beyond that length, the pendulum will stop rotating. The range in which your pendulum rotates is your personal wavelength. (The speed of rotation is also an indication of your vibrational level.)

5. Keep your thumb and index finger at the personal wavelength distance.

6. Put your non-dowsing/non-radiesthesia hand over a drug or supplement. By doing so, through your intention, you have linked yourself to the drug or supplement in question. You could be touching the drug or supplement, but you do not have to. Just having your hand over the container holding the drug or supplement connects you to the drug or supplement. You can prove this to yourself by testing the drug or supplement in the container, and by taking the drug out of the container and putting your hands on the drug or supplement.

7. Start the to-and-fro motion of the pendulum over your hand (which is also over the drug or supplement), and then observe the motion of the pendulum.

  - The pendulum will pull into a clockwise rotation if the drug or supplement is in resonance with you (i.e., good for you).
  - If the pendulum remains in a to-and-fro motion, the drug or supplement is neither good nor bad for you. It is not benefiting you, but it is also not harmful.
  - If the pendulum rotates counterclockwise, the drug or supplement is detrimental for your body (i.e., not in resonance with you).

You can also obtain an idea of dosage for drugs and supplements. You might vary the number of pills being tested (e.g., 1, 2, or 3 pills) and see what happens with pendulum rotation as you increase the number of pills.

Be aware that some drugs may not be good for you but your doctor wants a certain outcome in spite of the problem the drug may cause. In this case, you may want to talk to your doctor about side effects of the drug and then determine what you might want to do. It's somewhat similar to a dessert that may be very high in sugar, to which the pendulum will rotate counterclockwise. You may still decide to eat the dessert, knowing that you are not eating an excess of dessert every day.

8. Repeat this procedure with other drugs and supplements that you have brought out. You will now know which drugs and supplements are in resonance with you (beneficial).

Exercise: EMF Radiation (Electromagnetic Field)

1. Determine the string length of your personal wavelength. (See directions earlier in this chapter.)

2. Once you are at your personal wavelength, initiate the to-and-fro motion of the pendulum and connect to a source of EMF. At a minimum, you should test your cell phone, cordless phone, router for your WiFi, and computer using WiFi. You connect to

the phone, router, or computer by just putting your hands over the electronic device and seeing what happens to the rotation of the pendulum.

3. You will then see that EMF will cause a counterclockwise rotation. There are BioGeometry tools (strips) for the electronic devices that will harmonize the EMF.

About 3% of the population is very sensitive to EMF and will get quite sick from it. With the new 5G high levels of EMF that are being introduced for higher Internet speeds and a new generation of phones, protection is important. Currently, the Federal Communications Commission uses an outdated procedure to test for safety of EMF, in which if the EMF does not cause an increase in the temperature of a fluid, it is considered safe.

This is nonsense, since one needs to look at a biologic test, not something as crude as an increase in temperature. Scientists have proven that human cell receptors are affected by EMF, causing increases in calcium influx to cells, which initiates a cascade of detrimental effects. Fertility and other problems also occur. Metabolic pathways that cause free-radical formation occurs, which is likely the cause of cancers that won't show up until many years have passed. The good news is that you can protect yourself with things such as BioGeometry tools.

Stones (Minerals, Gems), Essential Oils, and Homeopathic Remedies

Minerals, gems, essential oils, and homeopathic remedies are often sold for their medicinal value. However, there is always the question of whether or not the particular stone or oil is truly beneficial *for the particular ailment you are trying to treat*. The best way to determine this is by testing, and the best test is with the personal wavelength method.

Follow the instructions given in the Personal Wavelength Exercise and see if the particular remedy will cause the pendulum to rotate clockwise for you (i.e., is in resonance with you).

You should also be aware that many minerals are irradiated to enhance the color of the crystals. From the perspective of giving beneficial energies, this is very bad and often gives off negative energies, causing the pendulum to rotate counterclockwise, making it unhealthy for you. Testing before you buy is essential.

There is much more you can learn about the use of the neutral pendulum if you are drawn to this subject. The BioGeometry Foundation Course teaches the use of the neutral pendulum and personal wavelength technique, plus much more: *https://www.biogeometry.ca/home* and *https://vesica.org/*

Having a Neutral Mindset

It's not just the pendulum that needs to be neutral. *You* do as well. You have to be in a certain state of mind—a neutral awareness—when you're dowsing with the pendulum. When you are using the pendulum, you're not thinking, "I want this pendulum to go clockwise" or "I want it to go counter-clockwise." You're just looking at something that the pendulum will give you information about, and being aware that you're looking at it and measuring its resonance energies.

You're not thinking anything beyond that. You're just letting the energy of your body and your right brain come down through your hand and cause the pendulum to go clockwise, counterclockwise, or back and forth so that it's the *energy* doing it, not your conscious mind.

Rather than thinking, "I'm going to make this a mental dowsing exercise, and I want the pendulum to go clockwise," you need to wait a second or so to let that energy flow from your mind into the pendulum and move in the direction it *should* move. You have to let the pendulum do its thing; and the best way to do this is to be neutral, to be in a state of awareness without thought. You don't want to be thinking, "I need to make this thing turn." You're just getting the pendulum into motion, and it will turn on its own.

BioGeometry, BG3, and Resonance

The field of BioGeometry works with resonance to detect and measure an energy quality called BG3 (or BioGeometry 3). We will be using BG3 measurements in a variety of ways in this book to help us understand energies and how they affect us. (BG3 is described in detail in *The Seeker and The Teacher of Light*.)

BG3 is an energy quality that is found in sacred spots around the world, and in the centers of structures (e.g., the center of a circle; an angle at the midpoint, such as 45 degrees for a 90-degree angle). BG3 has three qualities that can be measured separately, but it is actually a single quality of harmony and centering.

This is important because the presence of the quality of BG3 *produces* harmony, balance, and centering. The three qualities found in BG3 are (1) horizontal negative green, (2) the higher harmonic of gold, and (3) the higher harmonic of ultraviolet.

For example, a variety of problems are caused for people who are EMF-sensitive by microwave electromagnetic fields (EMFs)—migraines, disorientation, dizziness, nausea, and more. BioGeometry tools that produce BG3 can harmonize the negative effects of EMFs.

Dr. Ibrahim Karim, the founder of the field of BioGeometry, created a number of different pendulums that resonate with the BG3 qualities. A picture of two pendulums, BG16 and IKUP, are shown below. The pendulums use numbers and shapes to resonate with BG3. The IKUP pendulum can also be used as a neutral pendulum to find your personal wavelength. Both pendulums also emit harmonizing BG3 quality. Anytime BG3 is detected, the pendulum will rotate in a clockwise direction.

FIGURE 17. The BG16 pendulum (left) and the IKUP pendulum (right) are designed to easily and quickly detect BG3 through resonance with shape and numbers. The BG16 pendulum detects BG3, and the IKUP pendulum detects BG3 and can also be used as a neutral pendulum to find your personal wavelength. Both pendulums also emit BG3.

With a BG3 pendulum, whenever BG3 is encountered, the pendulum will rotate clockwise. If you purchase a BG3 pendulum from a BioGeometry distributor, this energy can be detected. As humans, we emit BG3, so a BG16 or IKUP pendulum will rotate clockwise over the back of your hand.

Measuring the Level of BG3

BioGeometry has a tool, a BG3 Ruler, to obtain a numerical value for the amount of BG3. 90-degree angles have a blocking effect on the transmission of the BG3 subtle energy quality. As a result, a row of such angles will block the BG3 energies. However, a stronger level of BG3 will result in a greater distance that the BG3 is able to travel. That is, a stronger BG3 will cross greater numbers of 90-degree angles before BG3 will no longer be detected. This procedure gives a relative BG3 value.

My purpose in describing the BG3 Ruler is to let you know that the science of BioGeometry and radiesthesia allows one to measure levels of subtle energies so one can quantitate results. The principles of the BG3 Ruler are described in Chapter 8 when there is discussion on the quantitation of BG3 levels.

3-6-9 Energies and Resonance

The 3-6-9 energy quality is the energy quality of the torus structure. As discussed earlier, the torus structure is the structure of creation. (This will be discussed in more detail in Chapter 6: "The Spiraling Energies," describing the findings of Walter Russell.) Structures and symbols related to the torus/3-6-9 include the Sanskrit word for aum, the bagua, Marko Rodin's Symbol of Enlightenment/Vortex Math (which mathematically forms the torus), and the yin yang symbol (also of the torus structure), as mentioned in Chapter 2. The details of the math showing the extrapolation of Vortex Math into the torus structure are described by Randy Powell in a YouTube video (*https://www.youtube. com/watch?v=gVN0vFAMfr4*).

All these structures are in resonance with BG3 (i.e., they all have BG3). In addition, if you create a pendulum with the key elements of 3-6-9 and the torus, then the pendulum will resonate with any of the 3-6-9 structures (i.e., torus, aum, yin yang, bagua, the numbers of 3-6-9 which show the energy rotation of the torus). These symbols are shown below:

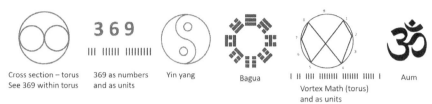

FIGURE 18. The above symbols have both BG3 and 3-6-9 subtle energy qualities, which can be detected by a BG16 pendulum and a 3-6-9 pendulum.

- The torus structure was activated by tracing the 6 and 9 in a clockwise manner.
- The numbers 369 are seen within the torus, and the numbers are activated by tracing the numbers in a clockwise manner.
- The numbers as units have BG3 and 3-6-9 qualities.
- The yin yang has been activated by tracing the inner clockwise/counterclockwise curve, which is also part of the torus structure.
- The bagua has geometric aspects of the isotropic vector matrix, which is derived from the torus.
- The vortex math drawing is activated by connecting the lines of 1-2-4-8-7-5-1, with the whole symbol being the mathematical equivalent of the torus.
- The numbers as units also have both BG3 and 3-6-9 qualities.
- The aum is self-activated and is basically the Sanskrit version of 369, which is the torus.

If you make a neutral pendulum, you can create a 3-6-9 pendulum by writing 3-6-9 on the pendulum as the numerals "3," "6," "9," drawn as shown in the figure below:

369

369 Drawn with CW
Direction for "6" and "9"
Shows BG3 and 3-6-9

FIGURE 19. The numbers 3, 6, 9 drawn with clockwise motion show BG3 and 3-6-9.

You can also create a 3-6-9 pendulum by writing the numerals as lines (III IIIIII IIIIIIIII):

3-6-9 as Units

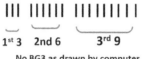

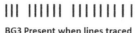

BG3 present if lines are drawn in order (1st 3, 2nd 6, 3rd 9 lines).

1st 3   2nd 6   3rd 9

No BG3 as drawn by computer

Lines drawn in reverse will not have BG3.

BG3 Present when lines traced over with pen in order shown

Cannot draw middle 6, then either 1st 3 and 3rd 9 – will not have BG3.

FIGURE 20. The numbers 3, 6, and 9 can be drawn as units (lines) to generate BG3 and 3-6-9 resonant energy qualities. The units (lines) must be drawn in order with space between the 3 lines, 6 lines, and 9 lines.

You can also create a 3-6-9 pendulum by drawing a cross-section of a torus structure, as shown in the figure below, with clockwise motion for drawing the two smaller circles representing 6 and 9.

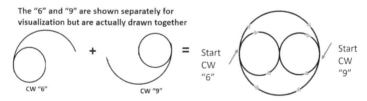

The "6" and "9" are shown separately for visualization but are actually drawn together

CW "6"   CW "9"

Start CW "6"   Start CW "9"

Follow the arrows to form the "6" and the "9", both drawn in one single motion. The 2 numbers form the torus cross section, with the "3" being part of the two smaller circles which are part of the "6" and "9". BG3 and 3-6-9 formed.

FIGURE 21. Cross-section of torus drawn to show 3-6-9 and BG3, using clockwise (CW) motion to draw entire structure.

The easiest way to create a 3-6-9 pendulum is simply draw 3 lines, 6 lines, and 9 lines. The lines must be drawn in sequence (1st 3 lines, 2nd 6 lines, 3rd 9 lines). The pendulum will resonate with the torus-related symbols shown in Figure 21, above. Although the symbols show both BG3 and 3-6-9, they are two different subtle energy qualities. This is easily demonstrated by putting a BG3 tool (a cube or L90 structure) with a 3-6-9 symbol; both subtle energies become canceled.

You can also purchase a pendulum from a store and write the numbers, draw a torus, or draw lines to represent the numbers. Write with a fine felt-tip pen or, better, etch this on with a carbide pen (see figures below). Through resonance, the pendulum will detect the torus-related structures described in Figure 21, above.

FIGURE 22. 3-6-9 pendulums made by writing 3-6-9 as lines on a stone pendulum, and on a neutral pendulum made from flour and salt. For greater permanence, the lines can be drawn with a carbide pen.

The same BG3 ruler that can measure the level of BG3 can also be used to measure the level of 3-6-9-resonant energies.

The 3-6-9 pendulum will rotate clockwise in the presence of 3-6-9 subtle energy qualities. As humans, we emit 3-6-9, and the pendulum will rotate clockwise over the back of the hand. The pendulum will rotate clockwise over any of the 3-6-9 symbols (see Figure 21 showing the various symbols).

Utility of BG3 Tools and 3-6-9 Symbols and Affirmations

BioGeometry has a wide variety of other tools, in addition to the pendulums, for bringing in BG3 to harmonize the environment or person. For example:

- BioSignatures engraved on a pendant or ring to bring in BG3 for the person and organs of the body.
- Strips to harmonize EMF.
- Cubes to bring BG3 to a room or house.
- Space harmonizers to harmonize a space.
- Archetype ruler to look for imbalances in the physical, vital, emotional, mental, and spiritual planes.

BioGeometry tools are tested for quality and proven to work exceptionally well.

There are a number of uses for 3-6-9. The symbols that give 3-6-9 subtle energy qualities will also bring in harmony. You just cannot combine the symbols together (i.e., put a 3-6-9 symbol on top of a cube), since that will nullify the effects of both. If the 3-6-9 symbols are at separate locations, there is no interference. The 3-6-9 symbols can be placed in certain geometric arrays to augment the amount of 3-6-9 energy, as measured with a BG3 ruler.

Affirmations are an important utility for 3-6-9. Affirmations have much greater impact when repeated 3 times, then 6 times, and 9 times. Many of Joachim Wippich's affirmations are recommended to be repeated 3, 6, and 9 times.

Summary

Direct experience with subtle energies allows you to experientially know that what is written here is true, and gives you further confirmation as to your I AM being. Using the tools for practical purposes (food, drugs, supplements) gives you a tool to know what is good for you rather than having to just guess.

PART III

Subtle Energies

In this section, I describe my experiences with seeing and feeling subtle energies. Various subtle energies and their properties are presented to familiarize you with qi, orgone, odic, and torsion energies. In addition, you are introduced to the works of Wilhelm Reich, Baron von Reichenbach, Nikolai Kozyrev, Klaus Volkamer, Ibrahim Karim, and Olga Strashun.

CHAPTER 4

My Experiences in Feeling Qi/ Subtle Energies

In this chapter, I thought you might find it interesting to read about some of my own interactions and practices with energies. I find that if I have experienced something, *others* also may have experienced similar phenomena but not registered this experience consciously, and therefore may have dismissed it as "not real." From my years of studying energies, I have become sensitive to many energies that, twenty years ago, I never knew existed.

Feeling Qi or Prana

In Eastern energy practices such as Tai Chi or Qigong (Qi Gong) used in exercise and martial arts, one works with the life-force energy, which is called *qi* or *chi*. The exercises are often used for enhancing health. In the Western world, qi or chi can be considered *subtle energy*. (There will be further discussions on subtle energies in the following chapter.) *Prana* is the word for subtle energy that is used in India.

Ever since my friend Guy Harriman, who taught energy yoga, introduced me to the energy known as *qi*, I have felt that energy. He brought me near some huge quartz crystals and had me "feel" the energy with my hands. The feeling was a "tingling" sensation in my hands any time I put them near the crystals. Guy explained that this was *prana*, or *qi* (chi). Guy and I, and his students, felt the same energy when we did exercises in the martial art Qi Gong, when we positioned our hands between the chest and the navel area and then built up energies between the palms of our hands. It seemed like a weight between the hands. We were able to build up quite a bit of qi within just minutes.

Later, when I learned about Tai Chi, I learned that these exercises were called "Dances with Balls of Energy." I found that any time I had my arms and hands facing each other, I would feel the same tingling sensation in my hands; and as I moved them as if in a dance, the energy would build up quite significantly.

Exercise: Feeling Qi

The ability to feel qi is something that develops as you practice feeling the energies. Here is an exercise for feeling qi:

1. Place your hands in front of you, with the palms facing each other, as if you are holding a basketball (about one foot apart).

2. Now move that invisible ball around (back and forth, up and down, in circular patterns) in a pretend Tai Chi exercise. Pay attention to the sensations you can feel between your hands. You might be surprised to feel a tingling sensation or the build-up of warmth between your palms.

3. Let the qi build as you play with that invisible ball of qi for about 5 minutes (i.e., have your palms facing each other about a foot apart).

Exercise: Seeing Qi or Prana

After finding out that I could *feel* qi, I wondered if I could also *see* the energy. I decided to try it out.

Looking between my hands and staring at a blank wall, I would see the lines (or rays) of energy between my hands. After practicing observing these lines, I noticed that when I looked at people, they would have rays above their heads. Often when you see paintings of saints, there are lines or rays above their heads.

In Qi Gong, one sends out prayers from the palms of the hand. I found that if I held up my hand in a moonlit room, I would see a beam of rays emanating from the palm. Just pointing a finger resulted in beams of rays coming out from the finger. I then found that when I looked at a quartz or amethyst crystal, I would see the same rays emanating from the tips of those crystals. At the gym, if I looked at a person doing the

StairMaster and with my eyes followed the rays above that person's head for about 4 or 5 feet as he or she went up the stairs, I would see a cloudy area that connected to the rays. Since I had heard about mental bodies, I assumed that was what this cloudy area might be. (In the literature about the various auric bodies surrounding us, there is the etheric body, which appears as a glow next to the skin; the astral or emotional body, further out; and beyond that, the mental body.)

Exercise: Seeing Lines or Rays between Your Hands in Front of a Light-Colored Wall

Here is an exercise to see if you can see the lines or rays between your hands:

1. Find a blank, light-colored wall (e.g., ivory or light beige, with no designs).

2. Hold your hands up with the palms facing each other about 8 inches apart, as if you are holding an invisible 8-inch ball.

3. Bring your hands to eye level, and look through the empty space between your palms.

4. Move both hands up and down simultaneously; or you can move one hand up a few inches, and keep the other hand steady. The movement allows you to tell the difference between seeing lines and no lines more clearly.

Exercise: Seeing Beams of Energy from Your Hands in the Dark

You can also try an experiment of seeing beams of energy that emit from your palms or fingers in a darkened room (no lights; just moonlight in the room).

1. Adjust your eyes to the dark, first.

2. Pretend that your palm is a flashlight. See if you can see a line of energy that is emitted by the palm.

Bringing Energy through the Body

Although I am not necessarily calling the energy described in this book "qi," it does have the same tingling sensation as the qi described above. In the meditations taught to me by Caroline Cory, the producer/director of *Superhuman*, we were asked to practice (a) bringing in Source energy from above the head, (b) bringing in Earth energy from deep within the Earth, and then (c) moving these energies through our bodies. I practiced this diligently for many months, and soon found the same tingling sensation within my head as when doing the qi practices. In the beginning of my studies, the energy feeling was limited to only a portion of my head. Nowadays, however, the tingling sensation runs throughout my head and into my body and hands.

Another practice I learned was the VELO Technique ("VELO" stands for Voluntary Energetic Longitudinal Oscillation), taught by the International Academy of Consciousness (IAC). IAC teaches how to have an Out of Body Experience (OBE). While in the hypnogogic state prior to having an OBE, a person experiences a lot of vibration in the body—which is why practicing getting to the *vibrational* state for an OBE is one technique for learning how to *have* an OBE.

In VELO, you consciously move your energies up and down the body, from the top of the head to the bottom of the feet. Then you make the rate of motion of energy very rapid, somewhat like playing with a toy paddle that has a rubber ball attached to a rubber band and seeing how the ball bounces back and forth on the paddle very rapidly. It is an excellent energy practice, one that helps you learn the feel of energy in the body. This practice was relatively easy for me, since I had been practicing bringing energy through my body with the Caroline Cory work. Because the VELO practice is best learned through the teachings of IAC, I am not including such an exercise here.

The I AM State

I find that as I repeat the affirmations taught to me by Joachim Wippich to bring myself to the I AM state, I often find myself connected to an energy sensation of tingling in my head and hands. If I pay attention, my whole body feels that way. As I am approaching the I AM state, checking my vibrational state with a pendulum or L-rod shows that the tool will rotate very rapidly. When I am fully in the I AM state, then there is stillness and I no longer feel the vibration. The pendulum also stops rotating.

For most people, just feeling the rapid rotation of the pendulum demonstrates the high vibrational state of I AM. Reaching the center of stillness when rotation ceases is the next, more advanced, state of being at the I AM essence level. For me, this higher level of the I AM state is produced by repeating the affirmations taught in this book 3 times, 6 times, and 9 times. Meditators typically reach the center of stillness after prolonged meditation, whereas Joachim's affirmations allow one to reach this state in minutes.

Exercise: Affirmations

Below is a set of affirmations I use to first bring myself into harmony, followed by affirmations I use to bring myself into the I AM state. Learning to check your I AM state (your vibrational level, as described in the previous chapter) is useful. I am repeating the Affirmation for Right Brain/Left Brain Harmonization here, so that you have the whole Affirmation in one place.

Affirmation: Right Brain / Left Brain Exercise for Harmony and Balance:

Hold your hands out to the side and then bring them together in front of your heart while saying each of the following affirmations:

Women:

"I AM inviting Every thought of my Left Brain and my Right Brain to join together with my heart I AM"

Men:

"I AM inviting Every thought of my Right Brain and my Left Brain to join together with my heart I AM"

This exercise can be done every morning or to bring yourself back into balance any time you think you might be out of balance. If you are dowsing, this should be your first step before dowsing.

Affirmation: Bringing Yourself to the I AM Level:

1. State this affirmation:

"I AM Everything I AM"

2. After making this statement, check for "Yes" or "No" using your dowsing tool. Even if your dowsing tool indicates "Yes," check what percentage is "Yes" by stating the following:

"My I AM is more than 10%."

3. If "Yes," continue as follows:

"My I AM is more than 20%."

and so on, until you reach 100%.

4. If your dowsing tool gives you a "No" response at a percentage lower than 100, then say (3 times total):

"I AM Rethinking Rethink I AM" "I AM Rethinking Rethink I AM"

"I AM Rethinking Rethink I AM"

Pause between your repetitions in order to absorb what you are saying.

1. Now recheck to see what percentage you are.

2. Repeat the above process until your "I AM" resonance reaches 100%.

Bringing Yourself to the I AM Level—Shortcut

The shortcut won't give you a percentage; its purpose is to simply check your vibrational level. Before you state the I AM affirmation, the pendulum will rotate at an average speed. As you reach the I AM level, the rate of rotation of the pendulum will become very rapid. You will then reach the level of quietness that usually is achieved after extended meditation, but in this case can be achieved rapidly by using the above method. The pendulum then stops, and all is quiet and at peace. When stating "I AM," breathe in with the "I" and breathe out with "AM."

The Energy Flow in Dowsing and Radiesthesia

When I am dowsing with a pendulum, there is energy flow that is due to the subconscious. The energy flows down the arm to the fingers holding the pendulum. When you are in harmony and at the I AM level, and you start the to-and-fro motion of the pendulum to overcome its inertia, the pendulum assumes its own life, as directed by the subconscious. If you are doing radiesthesia for testing purposes, the pendulum will move in the needed direction to give you an answer. There is typically a slight pause from the to-and-fro motion, and then the pull to go clockwise, counterclockwise, or to-and-fro occurs.

The Energy Flow During Healing Sessions

In addition to using Joachim's method of healing, where you bring in harmony and then bring up your vibration to the I AM level, I sometimes use other techniques that I have learned over the years. These include a combination of techniques, and are likely to include elements of other healers, such as Richard Gordon, Eric Pearl, Jane Katra, Reiki, Bill Bengston, Master Choa Kok Sui, Caroline Cory, and others.

- Richard Gordon is the author of *Quantum Touch*. In his earlier books, he used touch to initiate the connection/energy flow for healing. However, when I heard him speak a few years ago, he said that touch is not essential. He now heals with just intention for the energy flow.
- Eric Pearl is the author of *The Reconnection*. He uses an energetic connection to bring about healing.
- Jane Katra is a spiritual healer and uses a variety of spiritual connections to bring about healing.
- Bill Bengston wrote *The Energy Cure*. He teaches anyone to be able to heal, using a technique to enter into a healing "state of mind."
- Master Choa Kok Sui is the author of many books on *pranic healing*, in which energy is used for healing.
- Caroline Cory is a master healer and teaches a host of different techniques to help a person heal.

When I am acting as a healer and there are imbalances (e.g., pain) in the body of the person towards whom I'm directing healing energy, I will feel a tingling sensation in my hands as I move them over the body at a distance. This distance can be just inches away, or even as far away as a different city. Being a foot away from the person is ideal, since this proximity provides visual cues as to where you are in terms of the person you are working on; however, it is not necessary. You can use intention and visualization to know where you are relative to the person being healed.

I typically brush away the invisible excess "energies" causing the tingling sensation in my hands into a prepared solution of salt water, without touching the water itself. The bowl with salt water absorbs the energies. I view such energies as a form of "congestion" of energies; if they are not removed first, they can cause discomfort to the person you are working on. (This is well described in the teachings of Master Choa Kok Sui.) I then offer loving, harmonizing energies to the person being healed, who then can use those energies to self-heal.

I feel the tingling in my hands as I am offering the harmonizing energies. Once the person heals and their pain goes away, the tingling sensation in my hands ceases. This is the signal that the healing has been completed.

In such situations, one could say what's happening is that prana or qi is being given. I think of it as harmonizing energy that has BG3. I have also used Marcel Vogel's healing method using the Vogel crystals, which first are charged with energy. When I do so, I typically measure the amount of energy in the crystal with my BG3 Ruler so that I definitively know the crystal is fully charged and can be used to help bring in harmony for the person being healed.

It is important to note that the healer is not "doing the healing." The healer merely facilitates the person to heal him or herself. The harmony affirmations remind the body where the problem is located. The I AM affirmation brings the person to the vibrational level, where healing can occur. The qi or prana given by the healer does the same thing: a person's vibrational level is brought to a level where the body can do its work of healing.

Summary

I have experienced feeling and seeing subtle energies in a variety of ways. I hope that my description of these experiences in this chapter may be of help to you, to see if you might have similar experiences as those described in the various exercises. Of course, everyone is different and may experience energies in different ways than what I have described here. Some people may not see or feel anything—which is fine. We each have different sensitivities to energy, and those sensitivities may change over time as you work with energy.

CHAPTER 5

Subtle Energies—A Developing Story

Qi, Prana, and Earth Grid Lines: An Ancient Ancestry

Subtle energy practices are hardly new. Various practices of this kind were connected to ancient texts and tools. The *Wadj* and the *Ankh* are among such tools.

In ancient Egypt, the Wadj (pendulum) was a tool used for measuring as well as giving off subtle energies. (Pendulums have been discussed at length in this book and in *The Teacher and The Seeker of Light*.)

FIGURE 23. Wadj. See harmonizing BG3 and 3-6-9 from tip of pendulum

The *ankh* (see figure below) signifies life, viewed in more modern terms. But for the ancients, it undoubtedly served as an energy source for harmony and healing. From current BioGeometry testing, the ankh brings in BG3, a quality of harmony.

FIGURE 24. Ankh—Harmonizing BG3 and 3-6-9 are emitted from the top, bottom, and sides of the ankh.

Native cultures throughout antiquity have used subtle energies in a variety of ways. Priests, shamans, yogis, builders, and others have used a variety of tools for working with energies. Sometimes, such energy was called *qi* or *prana*, depending on the culture.

Practices using these energies included *healing* (*qi*, from the practice of Qi Gong) and *combat* (Kung Fu, Karate).

Cathedrals and sacred spots on earth were often situated on *ley lines*, which were located using vibratory rods or pendulums. These energy lines were known by different terminology in various parts of the world (e.g., as *Dragon lines* in China). In Spain, the famous 500-mile pilgrimage path along the Santiago de Compostada Camino lies along a ley line. The Mystery Schools of various cultures taught the use and understanding of such knowledge.

In addition to the major ley lines, there also are grid lines covering the earth, both on the ground and above ground. The *Hartmann lines*, discovered by Dr. Ernst Hartmann, run in a north-south and east-west direction, about 2 meters apart and 2 meters in height, forming a cube structure. These grid lines are about 20 cm wide, with wider grid lines (about twice the width) every 10 meters. The *Curry lines*, named after Dr. Manfried Curry, run diagonal to the Hartmann lines and are about 3.5 meters apart. Layered above the Hartmann grid system is the Benker Cube System, named after Anton Benker, about 10 x 10 x 10 meters square (thus the "cube" in its name). The grid lines carry information, which can be either beneficial or detrimental, depending on the direction of rotation of upward spiraling energies emanating from crossing associated with the lines.

BioGeometry practitioners often harmonize the major grid lines so that they carry beneficial BG3 rather than detrimental (vertical green) energies.

The Many Names of Subtle Energy

Many investigators have "rediscovered" subtle energy and have called it by a wide variety of names. Dr. Claude Swanson's book, *Life Force: The Scientific Basis*, Volume II, presents a table listing some of the main names that subtle energy has been called by various schools of study or investigators (see figure below). In this book, we will use Dr. William Tiller's name for it: *subtle energy*.

TABLE 1.2: SOME SYNONYMS FOR SUBTLE ENERGY

TERM:	SOURCE:
CHI, QI	CHINESE
KI	JAPANESE
PRANA	HINDU & TIBETAN YOGA
OD, ODYLE	VON REICHENBACH
ORGONE	WILHELM REICH
TIME DENSITY	NIKOLAI KOZYREV
TORSION	NIKOLAI KOZYREV
DELTRONS	WILLIAM TILLER
N-EMANATION	M. R. BLONDLOT
CHRONAL FIELD	A. I. VEINIK
VRIL	HINDU
BIOPLASMA	RUSSIAN, INYUSHIN
BIOGRAVITY	RUSSIAN, DUBROV
MANA	POLYNESIAN
ORENDO	IROQUOIS
WAKEN, WAKONDO	LAKOTA
BARAKA	NORTH AFRICA
PNEUMA	GREEK
MITOGENIC RAYS	GURVICH
ELAN VITALE	FRENCH (BERGSON)
LIFE FORCE	EUROPEAN
SUBTLE ENERGY	TILLER

FIGURE 25. Table of various names for subtle energy.

Modern Scientific Experiments with Subtle Energy

In more recent times, scientific experiments have been conducted using subtle energies with an impressive degree of success.

For example, the inhibition and healing of cancer cells has been shown in a variety of instances. In Dr. Claude Swanson's book, *Life Force: The Scientific Basis*, the author describes an experiment using qi to inhibit cancer-cell growth. Master Jixing Li, a Qi Gong master in California, inhibited cancer-cell growth in an incubator a thousand miles distant at the Penn State laboratory of Professor John Neely. In a different experiment at Dr. Joie Jones' laboratory, cells were given gamma radiation and treated with pranic healing, either locally or from a distance. This pranic healing was found to be effective at improving survival rates.

In another form of healing, Dr. William Bengston, in his book, *The Energy Cure*, showed that when the researchers created a certain positive state of mind as taught by Bengston and, on a daily basis, remained with the mice with cancer in that positive state for a defined period of time, the cancers healed within a month.

Dr. Swanson's book catalogs a variety of experiments that were carried out by Qi Gong masters. These experiments showed that the Qi Gong masters were capable of having significant increases in infared energy from their palms, which could cause fluctuations in the intensity of light in a laser beam and increase the magnetic field near the palm of the hand.

Odic Energy (OD)

The first modern Western scientist to investigate subtle energies was Baron Carl von Reichenbach, a German scientist who lived between 1788–1869. Because the term *subtle energy* was unknown at that time, he called the vital energetic force that he investigated *OD*, or *odic energy*, after the Norse god Odin. What von Reichenbach found was that certain persons, whom he called "sensitives," were able to see the odic energy. These people could see a visible glow from quartz crystals when their eyes were adapted to darkness.

Baron von Reichenbach did thousands of experiments concerning subtle energies. Here are some of his main observations and conclusions:

- Energy at the poles of magnets and crystals appear like smoke at the base and a flame at the tip. Heating weakens this effect.
- OD is found in sunlight and moonlight.
- OD flows through most materials—in the case of solids and insulators, at about 1 meter/second.
- OD behaves like light. It is refracted like light, and reflected by mirrors.
- This subtle energy has polarity, characterized as OD minus (-OD) and OD plus (+OD).

- The same polarities can be added to give a larger amount of that polarity, while opposite polarities cancel each other.
- OD minus appears blue and feels cool to the left hand but warm to the right hand.
- OD plus appears orange-red and is warm to the left hand but cool to the right hand.
- OD plus is attracted to electropositive elements (mostly metals such as copper or lead) on the periodic table, and OD minus is attracted to electronegative elements (e.g., oxygen, selenium). *Electronegativity* is a measure of an element's ability to attract electrons. *Electropositivity* is a measure of an element's ability to donate electrons.
- OD can be transferred from one substance to another; the energy dissipates over time.
- The heating of objects often forms positive OD; the cooling of objects often forms negative OD.

Orgone Energy

Dr. Wilhelm Reich (1897–1957) was an Austrian doctor of medicine and a psychoanalyst, a member of the second generation of analysts after Sigmund Freud. For the purpose of this chapter, we shall describe his work in the development of orgone, the orgone accumulator, and its use as a cloud buster to create rain. Let's look at his experiments, and then I will propose an alternative hypothesis as to what is happening.

In 1939, his lab assistant heated a sample of white beach sand over a bunsen burner until the sand glowed and swelled. What is significant in this is that no living organisms should have survived that amount of heat. Yet when the sand was placed in a solution of potassium chloride for 2 days, and then injected into a culture medium of egg medium in agar to see if anything would grow, a new yellow culture formed that had the shape of round vesicles (a small, round sack-like structure) with a diameter of 10-15 microns (a micron = one millionth of a meter in length). The vesicles glimmered with a blue light.

Dr. Reich called these vesicles "SAPA bions," for sand-packet bions. What was quite significant was that the bions had the ability to kill cancer cells. This property may resemble positive OD, which also has the ability to slow down or kill cancer cells.

The SAPA bions had the property of negatively affecting the space where they were stored. Holding a microscope slide with the bions caused the hand to become red and inflamed where the slide was handled. Something was radiating from the bions, since they also fogged photographic plates. The word "orgone" was coined to describe the energy being radiated from the bions.

To protect the environment, Dr. Reich built a box to house the bions, with metal on the inside and organic insulating material surrounding the metal. Reich then discovered that even when he removed the bion material, the orgone continued to accumulate in the box. In other words, orgone is ubiquitous in the atmosphere and other spaces. Thus, the concept of the *orgone accumulator* was born. The orgone blue-violet light, when magnified, appeared to pulsate, going from a smaller diameter to a larger diameter. Sometimes, the blue-violet intense light seemed to spiral in from the walls of the container. Orgone is apparently ubiquitous, since insulators and metal, when put together, result in the accumulation of orgone.

Dr. Reich also made cloud busters to create rain. A cloud buster consists of an array of hollow metal tubes connected to a body of water. The body of water would act as an absorber of orgone energy. Pulling orgone out of the atmosphere would cause the dissipation of clouds. It is believed that surface tension of the water in the cloud is decreased by pulling out orgone. This suggests that the moisture in the air, depending on wind conditions, could be localized to a particular area to cause rain by positioning the cloud buster at a number of strategic locations in the sky.

Orgonite is the term used nowadays for material made from plastic (such as polyester resin) containing metal pieces and some fragments of crystals (quartz). This acts as a natural accumulator of orgone.

3-6-9 and the Orgone-GANS Connection

Having some orgonite, I tested it for 3-6-9 subtle energy. It tested positive for 3-6-9 and negative for BG3. Putting 4 orgonite pieces in a rectangular or square shape produced BG3 and 3-6-9 within the perimeters of the orgonite. From this evidence, orgone can be considered as a form of GANS that will form plasma fields, as seen in the BG3 formed within the perimeter of the orgonite. The orgone accumulators and the orgonite also mean that the orgone is ubiquitous in our environment.

In Chapter 9, GANS is described as being monatomic matter with 3-6-9 energy. GANS produces fields containing BG3 and 3-6-9 subtle energies. From the 3-6-9 subtle energy, there is the connection with the torus structure. Orgone being ubiquitous, I went outside and tested for the presence of 3-6-9 and BG3 in the sky, and found that there *was* BG3 and 3-6-9 in the sky. Since I had learned that fields can be moved by intention, with intention, on a day with scattered clouds in the sky I invited orgone to accumulate in a portion of the sky. Clouds formed in the area where I had accumulated some orgone. Although this is not proof, it is an interesting anecdotal story.

Torsion Energy and Dr. Kozyrev's Findings

As a seeker into the nature of everything, I was gratified to come across the amazing work of the Russian scientist, Dr. Nikolai Kozyrev (summarized in Dr. Swanson's book). His scientific activities were revealed when the Iron Curtain came down and the work of key Russian scientists became known.

Dr. Kozyrev was troubled by the very same concepts that had troubled me as a college student—that everything in the universe runs down and dissipates. He postulated that there is a *spiraling* energy, a *torsion* energy, which balances the universe—that for everything that dissipates due to entropy, there is a corresponding generative/creative effect that occurs. This phenomenon can be called *negentropy* (a word that was not in the lexicon back in the 1960s, when I was learning these concepts). Thus, there is a negative and positive component to torsion energy—one that is involved with creation/generation, and another that is involved with radiation/dissipation.

In Dr. Kozyrev's model, increasing entropy (loss of order) results in the emitting of right-handed torsion that occurs elsewhere, resulting in *negentropy*. Decreasing entropy (increase in order) leads to the emitting of left-handed torsion that occurs elsewhere, resulting in *entropy*.

Dr. Kozyrev then devised extremely sensitive balances, gauges, and other instruments to measure the changes in very controlled experiments when something dissipates (entropy), as well as when something becomes more organized (creation). Many of these experiments and their instruments are very well described in Dr. Swanson's book. For entropy, he studied such processes as acetone evaporating, plants dying, ice melting, certain chemical reactions, and sugar dissolving. For negentropy, he looked at experiments such as stretching rubber bands, plants growing, ice freezing, and compressed springs.

Dr. Kozyrev carried out many types of experiments, including specially designed telescopes using torsion detectors. He found that he could detect the positions of stars and planets at the current time, past time, and a future time. Therefore, the torsion waves are not based on the limitations of the speed of light. Thus, what is happening is not dependent on our standard concepts of time based on the speed of light. The concept of balancing entropy with negentropy anywhere in the universe and doing it instantaneously is possible when speed of light is no longer an obstacle.

The Subtle Energies of BioGeometry and Radiesthesia

Much of this book is based on the study of subtle energies, as derived from the work of Dr. Ibrahim Karim and his development of the science of BioGeometry, based on the centering, harmonizing qualities of BG3. Dr. Robert Gilbert at the Vesica Institute has been investigating the subtle energy qualities of the outer bands of the vibrational energy spectrum. Below is a figure showing the outer bands of the energy spectrum.

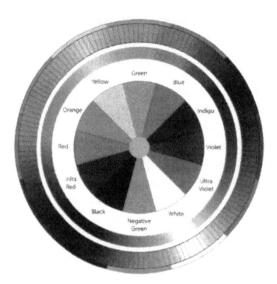

FIGURE 26. The Vibrational Spectrum. Subtle energies are associated with each color, which can be sub-divided into 144 sub-bands. The bands can be further sub-divided into beneficial horizontal energies and typically non-beneficial vertical bands.

Various minerals can often be characterized by the color associated with them. Examples include black, which typically has the capacity to absorb toxic radiation; infra-red, which brings warming and often can be used for acute pain relief; and red for activating, vitalizing. Each color has its effects. Using the personal wavelength techniques, you can determine which essential oil or mineral will be beneficial for various ailments.

The techniques of BioGeometry and radiesthesia to measure BG3 and personal wavelength have been primary tools in many of the experiments in this book. The principles of measuring the 3-6-9 subtle energy are also based on radiesthesia.

My Experiments

Based on my understanding of BioGeometry and radiesthesia, I decided to check out my developing hypothesis that clockwise rotation of a pendulum represents negentropy and that counterclockwise rotation represents entropy. Reviewing the list of experiments carried out by Dr. Kozyrev, I repeated them with just a pendulum as the measurement tool.

I duplicated Dr. Kozyrev's experiments and showed that entropy (the breakdown and winding down of the physical universe) is accompanied by a counterclockwise rotation of the pendulum, whereas what could be called negentropy (what is alive and growing) exhibits a clockwise spin of the pendulum. My experiments matched the results of Dr. Kozyrev, who carried out his experiments with supersensitive balances, gauges, and other tools.

Experiments You Can Do to Prove the Spirals of Creation and Decay

You can do these kinds of experiments yourself—such as evaporating acetone, freezing water, stretching a spring, and testing living matter. They are simple to perform. Then you will know for yourself, through experience, that creation (negentropy) is centripetal and clockwise, and decay (entropy) is centrifugal and counterclockwise.

Exercise: Entropy and Negentropy Pendulum Spins

Testing the spin of the pendulum with entropy, using evaporating acetone (typically available from hardware stores; if you don't have acetone, you can use fingernail-polish remover).

1. Using a neutral pendulum, hold the string at about 1.5–2 inches from the weight on the pendulum.

2. Over the closed bottle of acetone, start a to-and-fro motion of a neutral pendulum. The pendulum will remain in a to-and-fro motion.

3. Now open the bottle of acetone and pour a little of the acetone into an open vessel (e.g., a cup, bowl, or the cap of the acetone bottle).

4. Start the to-and-fro motion of the pendulum.

5. The evaporating acetone will cause the pendulum to rotate counterclockwise (decay, entropy).

Testing the spin of the pendulum with negentropy, using stretched rubber band:

1. Using a neutral pendulum, hold the string at about 1.5–2 inches from the weight on the pendulum.

2. Place a rubber band, in its relaxed position, on a surface.

3. Hold the pendulum over the rubber band.

4. Start a to-and-fro motion. The pendulum will remain in the to-and-fro motion.

5. Put one end of the rubber band on any stationary protrusion (nail, screw, etc.) and stretch it out.

6. Start the to-and-fro motion of the pendulum over the stretched rubber band.

7. The stretched rubber band will cause the pendulum to rotate clockwise (negentropy).

The Subtle Matter Work of Dr. Klaus Volkamer

An important addition to the discussion on subtle matter is the scientific work of Dr. Klaus Volkamer, author of *Discovery of Subtle Matter*. Through 30 years of research using ultrasensitive detectors (e.g., balances), Dr. Volkamer has proven the existence of subtle matter.

This subtle matter is associated with *all* matter, especially at "phase boundaries" of matter (surface areas where density gradients exist), which are easier to measure at interfaces of solid-solid, solid-liquid, solid-gas or liquid-gas interfaces. Sensitive balances that were able to measure accurately in the microgram range were able to measure the absorption of subtle energies at these boundaries.

Water can change due to subtle matter. Different weights when entropy occurs could be measured. The delicate instruments were able to make measurements with chemical, physical, and biological systems. The subtle matter also had properties that are known in particle physics. The subtle matter also had a field-like quality, extending at least decimeters from the macroscopic mass.

The subtle matter had either positive (weight-increasing) signs or negative (weight-decreasing) signs—properties that are unknown in gross-particle physics. Results of these experiments indicate that both forms of subtle matter, both weight-increasing and weight-decreasing, can be absorbed by celestial bodies, yielding significant gravitational anomalies to these bodies. Weight changes of a silver-plated sample varied with moon-rise to moon-set, and also with the stars behind the moon. The only explanation for the data was due to the absorption of a form of invisible subtle matter due to the moon and stars.

Dr. Volkamer observed that there are two types of subtle matter. One can be assigned a positive sign, which has a gravitational (weight-increasing) effect and an entropic (health-damaging) effect. Subtle matter that has been assigned a negative sign has anti-gravitational (weight-decreasing) effects, and negentropic or syntropic (supporting-health) effects.

Dr. Volkamer also says that subtle matter may necessitate an extension of thermodynamics by adding a fourth law—i.e., a law of "negentropy" or "syntropy"—that can counteract the second law of thermodynamics (the law of increasing entropy in spontaneous processes of gross matter).

These are the salient details:

- Subtle matter is absorbed at phase boundaries of gross-matter materials. The result of this absorption is a measurable and significant change of weight of the gross-matter material.
- Subtle matter also shows a gravitational interaction with gross matter and with itself. This results in gravitational anomalies of weight change.
- Subtle matter also appears to have a weak electromagnetic interaction with gross matter. However, the strength of this interaction is so weak that the human eye is not able to detect subtle matter.
- Subtle matter also shows a tendency to form clusters in the same way that atoms can form molecules.

This discussion of Dr. Volkamer's work is intended to lend further scientific credence to the field of subtle energy and subtle matter.

Subtle Energy Principles of Dr. Olga Strashun

Dr. Olga Strashun has been studying subtle energies for the past 30 years. She is a co-founder of the Strashun Institute, a Subtle Energy researcher, practitioner, and author of the book, *Subtle Energy: Information to Enhance, Guide, and Heal.* She was an MD in Russia when she started her training at the Subtle Energy laboratory founded by the USSR Ministry of Space. In 1989, Dr. Strashun emigrated to Canada with her family, where she continued her investigation in this field.

I listened to her presentation at the 2021 Joint Conference of the Parapsychological Association and the Society for Scientific Exploration. I found her 10 Principles of Subtle Energy compelling enough to present those principles below. My own findings reflect what Dr. Strashun has observed about subtle energies.

Principle 1: On the deepest level of reality, we are all made from the same matter and interconnected.

Principle 2: The Subtle Energy field is neither electromagnetic nor gravitational in nature.

Principle 3: Subtle Energy carries information. *Our thoughts are the purest form of Subtle Energy.* Thoughts are a force affecting matter.

Principle 4: Everything in the world vibrates. The quality of the vibrations is important information.

Principle 5: All animate and inanimate objects have a dual nature: a visible aspect and a vibrational aspect. All things in the world perform just as a vibrational form.

Principle 6: The Universe is like a giant "hologram." Everything in the world can be expressed through its vibrational form.

Principle 7: Information in the form of vibrations never disappears; past, present, and future exist in each and every point of space and time.

Principle 8: By achieving resonance with an object's vibrations, the vibrations automatically enlarge and can even be changed.

Principle 9: Concentration, like a laser beam, can retrieve any wave form of any subject/object of our interest and turn it into a holographic image we can easily work with.

All these principles from the domain of vibrations, as codified by Dr. Strashun, reflect the eternal and infinite nature of subtle energy, which concerns everything in the world.

In her book, Dr. Strashun describes how she works with subtle energies to gather valuable information on the medical condition of her patients. She gives practical information on how to develop the capabilities of working with subtle energies to achieve valuable information in many areas of life.

Summary

There is a vast amount of information on the subject of subtle energies. Dr. Claude Swanson has a list of 24 names for subtle energy in his book, *Life Force: The Scientific Basis* (Table 1.2). This chapter has introduced you to some of the main names given to subtle energy: *qi, prana, OD/ odic energy, orgone, torsion energy,* and *subtle energy.* In this book, the main term we use to describe the energies is *subtle energy.*

It also has presented information about qi, prana, the properties of Baron von Reichenbach's odic energy, Wilhelm Reich's orgone energy, Dr. Nikolai Kozyrev's torsion energy, and Dr. Klaus Volkamer's proof of subtle matter.

Dr. Ibrahim Karim has made significant advances in understanding subtle energies with his development of BioGeometry and discovery of BG3.

Dr. Olga Strashun has created a list of principles on subtle energy, based upon her many years of working in this field.

PART IV

Spiraling Energies

The centripetal spiral is the basis of matter formation. Spiraling energies are part of our vibrational essence.

The creation of matter is based on the spiraling motion of energies. This motion is the basis of the torus.

Nassim Haramein, Walter Russell, the scientists described in Foster Gamble's *Thrive* documentary, the math of Elizabeth Rauscher, the math of Marko Rodin and Randy Powell, all emphasize the spiraling motion and the torus structure. I like the descriptions of Walter Russell, the early pioneer and enlightened individual who gave us a solid glimpse into the nature of creation. There is also the body of work from Dr. Nikolai Kozyrev, who described the spiraling torsion energies and his exact measurements of those energies. Dan Winter developed the "phase conjugate" concept of the golden ratio spiral, which gives further evidence that the centripetal spiral is the basis of matter formation.

In college, when I was studying the laws of thermodynamics, I learned that one of the key laws had to do with the concept of *entropy*. The law simply states that everything in the universe winds down (dissipates); that entropy is the natural flow of everything. I remember being disappointed with this law, since it makes life and everything we do less important—that is, "it is all going to dissipate anyway." However, this law of thermodynamics does not hold up when we consider the spiraling energies of creation.

In subsequent chapters in Part IV, we will see that spiraling energies are the nature of who we are—part of our vibrational essence.

CHAPTER 6

The Source of Spirals

*Since everything is light, it is the motion of light that
is spiraling inward for creation of matter. In other
words, we, and everything else, are created by the
motion of light. Different manifestations of matter are
simply the different relative motions-in-equilibrium
of light formed by varying conditions in their
spiral journey.*

This chapter will pull together the directions of the spiraling energies
to indicate the validity of the fundamental nature of creation.

In Chapter 3, "Resonance, Subtle Energies, and Tools for Their
Detection," we saw that the clockwise (CW) motion of the pendulum
is the basis of personal wavelength, and that anything that resonates
with us revolves clockwise, whereas anything detrimental to us revolves
counterclockwise (CCW). In this chapter, I will review Walter Russell's
illuminations in more detail. In addition, I will highlight the excellent
work of Dr. Nikolai Kozyrev, the Russian scientist who taught us about
torsion energy: that the energy of anything that is in the process of creation has a clockwise spin to the torsion energy, while anything that is
in the process of decay has a counterclockwise torsion spin.

Walter Russell and the Spiraling Energies of Matter Creation

When the nature of creation was revealed to Walter Russell, he imme-
diately understood that matter is created by the centripetal clockwise
spiraling of light to compress light into matter. He also understood that

the creation of matter was half of a cycle, the other half being the dissolution of matter by a centrifugal counterclockwise spiraling of light to dissipate matter. Since it is a cycle or wave, the dissipated light will repeat the cycle, to again be compressed into matter.

These few sentences offer a fairly simplistic view of creation, but they do give the gist of the story that has been told by Walter Russell in his various volumes of books to give a more complete scientific view of matter creation.

In this story, clockwise rotation is always associated with the *generative* part of the cycle of matter creation, while counterclockwise rotation is always associated with the *radiative, dissipation* aspect of matter decay. In Chapter 3, "Resonance, Subtle Energies, and Tools for Their Detection" addressing vibrational energy and radiesthesia, we learned that anything that is in resonance with us (i.e., is life-promoting) causes the pendulum to rotate in a clockwise manner. Conversely, if something is not in resonance with us (i.e., is not life-promoting, or is even harmful to us), the pendulum rotates in a counterclockwise direction.

The way Walter Russell puts it is that *everything is light*. Thus, it is the motion of light that is spiraling inward for creation of matter. *In other words, we, and everything else, are created by the motion of light.* The figure below from Russell's book shows the overall dynamics.

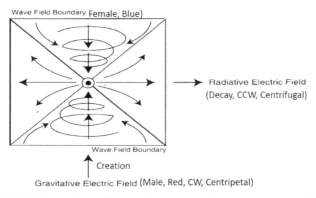

FIGURE 27. Formation of Matter–Creation as the vortex motion of light.

Creation is a wave that encompasses both creation and decay. Creation is the generative, electric effect of light (the red/male aspect) spiraling inward in a clockwise centripetal motion. Simultaneously, the blue female light also spirals inward in a clockwise centripetal motion.

Both the two cones of light are spiraling inward in a clockwise direction. As they move in, they go faster and faster as the light moves towards the center. An opposite clockwise spiral comes in from the other side to meet the first clockwise spiral, one from the bottom and the other from the top. (In the next figure, Walter Russell describes the two cones of light as meeting at the North position, rather than calling it "top" and "bottom.")

The conjunction and vortex of the spirals results in an illusion of solidness, which Russell calls motion-in-equilibrium—that is, matter. It's like two cars going head-on to each other: there is the illusion they have come to a stop (or slowed down) because they're coming together. That's the motion-in-equilibrium.

In terms of shape, matter moves from an elliptical shape further from the center, and becomes spherical when the cones of light meet in the center of the vortex.

The decay of matter then commences with the counterclockwise radiative, magnetic, centrifugal motion of light to dissipate matter.

The cycle is continuous, creating waves of creation and dissipation. Walter Russell's typical illustration of creation is shown in the figure below.

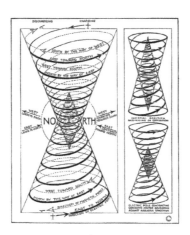

FIGURE 28. Another representation from Walter Russell, showing the vortex motion of light to form matter.

In the creation process, there are two spiraling sets of light: red (male) and blue (female). This creates the first concept of polarity.

The inward spiraling of light can be described as *gravitative or electric,* and the outward spiraling of light can be described as *radiative or magnetic.*

Another description would be *creation and decay.*

Motion-in-Equilibrium

The concept of how matter is formed from spiraling energies may be difficult to conceptualize. This section will describe Walter Russell's concept of converting motion to matter.

What happens is that the compressing spirals of light move inward toward each other. The movement can be divided into 4 steps, or *octaves.* With the 2 spirals of light, there are 8 octaves; and if you also count the beginnings of the octaves, there are 10 octaves. The motion of spiraling lights interacts in the compression process and helps give the illusion of stability in the motion, lending matter the appearance of stability. Russell calls this *motion-in-equilibrium.*

At each of the octaves, different elements form, with carbon being the highest-compression state of matter. Thus, different manifestations of matter are simply the different relative motions-in-equilibrium of light formed by varying conditions in their spiral journey.

The two spiraling sets of light account for the double torus as the basis of creation. This becomes even clearer when you study the work of Nassim Haramein, working with the math of Elizabeth Rauscher.

The spirals of light do not stop spiraling. After the highest-compression level, they continue their path to complete their radiative disintegration. This is the beginning stage of entropy, or dissipation of matter.

Some elements are more advanced in the dissipation process. These are the *radioactive* elements. Walter Russell created a whole periodic table based on where the elements are in their octave position of motion in the spiral. The outward spiral from the center is the *centrifugal counterclockwise motion of dissolution,* or *radiation.* The inward spiral towards the center is the *centripetal clockwise generative* part of the cycle of creation.

The spiraling of light inward and outward is a cycle, or wave. At the completion of the radiative decay, which Russell often describes as "magnetic inertia," the new cycle begins again with compression of light in inertia to create matter.

The Directions of the Journey of Light

Russell also uses a different, directional terminology to describe the journey of light. *North* is the direction of the highest compression, at the tip of the cone of the spiraling light. This is the centripetal, clockwise, gravitative, electric direction of creation. *South* is the direction of decompression, or radiative/magnetic breakdown.

As the two spiraling cones of light meet, the counterclockwise, radiative, magnetic direction of breakdown occurs. The motion-in-equilibrium occurs, giving the illusion of stability. The spiraling journey of light ends with full radiative dissipation.

But since this is a wave, the dissipated South begins its compression journey again to the North.

Another way to view Russell's diagram is to consider the two centripetal spirals of light as being the north end of a magnet. The huge momentum of the spiraling light (speed of light) keeps the light moving into the tight, compressed cone. At the same time, another cone of light is spiraling in from the opposite direction, under the same momentum.

An ordinary bar magnet illustrates this process. When the two ends of the magnet come together, if they are of the same polarity (i.e., north), a huge magnetic repulsion occurs. Similarly, with the movement of light, the revolutions of the light will seem to slow down, and the two spirals of light will appear to be stable (motion-in-equilibrium). This gives the illusion of solid matter.

The different types of matter (elements) will form on different parts of the spiral, creating the elements of the periodic chart. Carbon, for example, is at the apex of the cone and is of the highest density.

As the spirals of light merge and go through each other, the magnetic, radiative forces break down the matter. Thus, after full penetration of one spiral with the other spiral, matter at the outer edge of the spirals will be at the position of greatest breakdown (i.e., decay). These are the radioactive substances, which spontaneously undergo radiation breakdown. As the two spirals meet at the center cones, the magnetic repulsion forces create the counterclockwise centrifugal spiraling motion of light, leading to the dissipation of matter.

The path of this process, shown as a wave, appears in the figure below. Here, using Russell's alternative directional terminology, *anode* is the cone center, or North, and *cathode* is South.

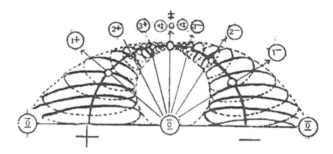

FIGURE 29. Walter Russell's wave view of the matter-creation process. In the center, the two cones of male and female light spiral in to meet at the cone apex of highest density. The spirals show the different octaves of compression, going from zero to +1, +2, +3, +4, and 0 at the conjunction of the cones.

For those wanting to read more about Walter Russell's many works, I would recommend starting with his biography to get the summary of who he was and the general scope of his work: *The Man Who Tapped the Secrets of the Universe,* by Glenn Clark. To learn how Russell acquired his knowledge in his downloaded message, read the two volumes of his book, *The Message of the Divine Illiad.* To understand his science, read *The Universal One* and *In the Wave Lies the Secret of Creation.* To become the "Cosmic Man," study the 10 books in his *Universal Law, Natural Science & Living Philosophy: A Home Study Course.* The best paintings depicting the waves of creation are featured in Russell's book, *In the Wave Lies the Secret of Creation,* as well as in *The Secret of Light.*

Dan Winter and the Golden Ratio Spiral

Dan Winter—scientist, researcher, and sacred geometry expert—has developed a large body of work showing the importance of the spiral.

This work is summarized in his book, *Fractal Conjugate Space & Time: Cause of Negentropy, Gravity and Perception: Conjuring Life: "The Fractal Shape of TIME" Geometric Origins of Biologic Negentropy...* (*http://www.fractalfield.com/conjugategravity*). His work focuses on the golden-ratio spiral's ability to compress the ether to explain many observations, such as the basis for life (negentropy), photosynthesis, gravity, and time.

For purposes of this chapter, we will focus on the nature of the spiral to compress the ether. This is analogous to Walter Russell's centripetal spiraling compression of light for creation of matter. Winter gives another name to ether—"charge"—but acknowledges that others have called it "ether."

Golden Ratio (φ): Phi=1.618

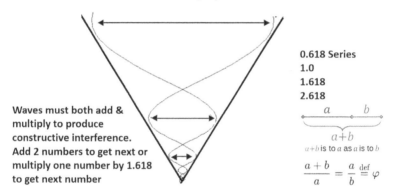

FIGURE 30. Dan Winter has found that compression waves that follow the golden ratio cause constructive interference, which results in other waves following a similar pattern (***phase conjugate***) to compress in the same manner. He defines the compression pattern as ***implosion***. Spirals are activated by tracing them with a pencil. When activated, BG3 and 3-6-9 are found to be associated with the spiral curve. You can prove this by following the spiral with a BG16 or a 3-6-9 pendulum (the tools to detect BG3 and 3-6-9).

The Schumann Resonance Frequencies

A fascinating application of the golden ratio compression of waves is the *Schumann resonance frequencies*. Otto Schumann, a 20th-century German physicist, discovered that there are energy peaks in the spectrum of extremely low frequencies (ELF), which are the effect of global electromagnetic resonances that have been excited by lightning discharges. When this occurs, the Earth forms a cavity between its surface and the ionosphere. Frequency peaks are found around 3, 8, 14, 20, and 26 Hertz.

The Schumann resonance frequencies are akin to the frequency of the Earth. We are constantly exposed to those frequencies. Evidence continues to build that much of life is attuned to those frequencies. Astronauts in space become disoriented unless these frequencies are given to them artificially. In fact, I have a Pulsed Electromagnetic Frequency (PEMF) mat that I use because of the known benefits of getting enough of the Schumann frequencies. The frequencies improve cellular health, bring more oxygen to cells, and help remove toxins.

It turns out that the Schumann frequencies follow the golden ratio spiral, as shown in Figure 31.

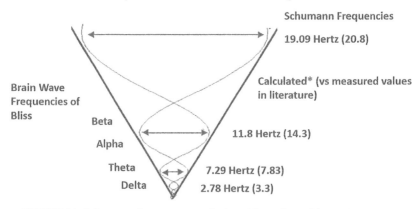

FIGURE 31. Schumann frequencies calculated from the golden ratio are very close to the frequencies measured by Otto Schumann. These frequencies are very similar to brain-wave frequencies, especially those found in the bliss state. The spirals are activated by being traced with a pencil to make it a continuous spiral. When activated, BG3 and 3-6-9 are found to be associated with the spirals.

A question arose for me: If the golden ratio is the basis of the centripetal vortex compression for creation and is in the torus form, if the frequencies were expressed as units, would they have 3-6-9 and BG3? This indeed turned out to be the case—see Figure 32, below.

Schumann Frequencies: 3-6-9 and BG3

Frequencies Translated to Units	BG3 Level*	3-6-9 Level*	
19.09 ‖‖‖‖‖‖‖‖‖‖‖‖‖‖	Paper facing up, print on top of paper		
11.8 ‖‖‖‖‖‖‖‖‖			
7.29 ‖‖‖‖‖‖	300	0	
2.78 ‖‖			
19.09	Photocopy of back side of paper, cannot see units but 3-6-9 present		
11.8			
7.29	300	300	
2.78			

*Different types of symbols can sometimes show BG3 or 3-6-9 on top of paper, sometimes on reverse side of paper and often must be determined empirically or with aid of Turenne Pendulum

FIGURE 32. The Schumann frequencies–calculated from the golden ratio, when converted to units–are in resonance with BG3 and 3-6-9. The lines are activated by being traced with a pencil, so they can be considered a unit.

Another way to show the presence of BG3 and 3-6-9 associated with the golden ratio spirals is to convert the golden ratio sequence into lengths on a ruler, as shown in the figure below.

Golden Ratio Sequence: Line length as Measure of Number

0.618	
1.000	BG3 and 3-6-9
1.618	of 300 on BG3
2.618	Ruler

FIGURE 33. Golden ratio sequence shown as lengths reveals both BG3 and 3-6-9.

In connection with the work of Dan Winter, it has been found that the Schumann frequencies are closely related to the brain-waves frequencies, especially those involved in bliss. The alpha, beta, delta, and theta frequencies of the brain are also involved with our different states of consciousness.

We have learned that the torus structure, with its spiraling vortex motion, is the basis of the creation of matter. I have always assumed that spiraling motion was related to the Fibonacci sequence and its resultant spiral. (The Fibonacci sequence is a close approximation of the spiral created by the golden-ratio numbers.) As it turns out, Dan Winter has shown that the spiral is actually *determined* by the golden ratio.

We have learned that the 3-6-9 resonances are associated with the torus, and that the spiraling motion of light is associated with creation. The spirals associated with the golden ratio, as exemplified by the calculated Schumann frequencies with its 3-6-9 subtle energies, give further evidence of the fundamental nature of the torus in the process of creation. Resonance with these energies was demonstrated in three ways: (1) as the spiral itself, (2) as the peak values of the spirals, represented as number of units as lines, and (3) as lengths on a ruler.

We Live in a World of Octaves

Another way to view this information is to realize that we live in a world of octaves. These octaves are based on the golden ratio, phi. Our senses of sight, taste, touch, smell, and hearing are tuned to different octaves of the golden ratio. Thus, it is not surprising that we can sense the different octaves of subtle energies in addition to the physical ones that we see, hear, taste, and smell. Our vibrational energies are tuned to the golden ratio.

Dan Winter made a number of very important observations regarding the spiral and the nature of compression of light (ether/charge). His book, *Fractal Conjugate Space & Time* (etc.), delves into how all of the following are related to the spiraling compression of ether ("light," in Walter Russell's terminology): (1) the dodecahedron shape of Earth; (2) the life force and negenetropy; and (3) the basis of gravity, color, and photosynthesis. Winter discusses these and other concepts as *phase conjugation of spiraling charge and their compression* (implosion), to explain the key concepts mentioned. I thought it important to explain his concepts of the spirals in connection with creation and the subtle energies that are described in this book.

Dr. Kozyrev and the Spiraling Torsion Energy

We discussed the work of the astronomer and astrophysicist Dr. Nikolai Kozyrev and the spiraling torsion energy in Chapter 5, "Subtle Energies—A Developing Story." He rightfully believed that thermodynamics could not be correct in predicting that the ultimate decay of everything was the destiny of the universe. Assuming that there had to be a net creative process that would balance any decay process, he postulated the existence of spiraling torsion energy to counter entropy. This is analogous to the concept of the creation-and-decay cycle described by Walter Russell.

When measured by instruments, the experiments of Dr. Kozyrev demonstrated that the creative processes resulted in a clockwise spiral, and that the decay processes resulted in a counterclockwise spiral. We demonstrated this with the simple use of the direction of rotation of the pendulum. Clockwise rotation is creative (negentropy), and counterclockwise is dissociation or decay (entropy).

The Spiral in Radiesthesia and BioGeometry

We have learned that when we measure with the neutral pendulum and look at our personal wavelength, if something is *not* in resonance with us (i.e., is detrimental) the pendulum will rotate counterclockwise, whereas if something *is* in resonance with us (i.e., is good for us) then the pendulum will rotate clockwise.

This means that if something is healthy for us, it contributes to negentropy, whereas if something will make us ill, it contributes to entropy. Thus, the spiraling directions for Walter Russell's creation-decay cycle match Dr. Kozyrev's entropy-negentropy torsion energies.

Similarly, with the BG3-detecting pendulums, detrimental things cause counterclockwise rotation, and beneficial things cause clockwise rotation.

All this matches the spiraling directions that are observed in radiesthesia and BioGeometry.

The Spiral in the Vibrational Level of I AM

When we check our vibrational level with a pendulum, as discussed in Chapter 3 and in *The Seeker and The Teacher of Light*, the pendulum rotates clockwise. The speed of rotation gets very rapid as we move ourselves toward the I AM level, using the affirmations taught by Joachim Wippich. This is further evidence of the generative or creation portion of us. We are alive, and therefore rotation is clockwise for pendulums or L-rods.

Summary

The spiral is an integral part of creation. Clockwise spin is indicative of the creation part of the cycle, and counterclockwise spin is indicative of the dissipation/decay part of the cycle. This has been shown in Walter Russell's creation model, in Dr. Dan Winter's discoveries, in Dr. Nikolai Kozyrev's torsion-energy model of entropy and negentropy, and in the pendulum rotations of radiesthesia and BioGeometry. When we test our vibrational level and we are in resonance, the spin is clockwise. Dr. Dan Winter's work shows that the shape of the spiral is related to the golden ratio (phi). Numbers created from the spiral also demonstrate the presence of BG3 and 3-6-9.

CHAPTER 7

The Nature of Vibration Is in the Spiral and in Our Nature

Raising the Vibrational Level

"Raising the vibrational level" is something that I, as well as many other people, often talk about. *The Seeker and The Teacher of Light* speaks of the high vibrational level we are at when we realize that our essence is I AM. And as we have learned, the pendulum rotates clockwise very rapidly when that occurs.

For most of us, when we talk about vibration, we think of something like a string or a rubber band moving back and forth very rapidly. But nature does not work that way. Vibration is in the *spiral*.

What we have learned from studying the findings of Walter Russell is that everything is motion, and that the motion takes the form of a spiral. Thus, what is spiraling is light, or electromagnetic energy. And as we have learned, there are two sets of light (male and female, or red and blue) that are spiraling, and that they are spiraling toward each other.

Let's examine the nature of the spiral.

The Nature of the Spiral

At the beginning of the spiral, its diameter is larger, and so the revolution of light has a larger diameter. As the spiral moves to its apex, the diameter becomes smaller and the rotation of the light becomes faster. It's like when you pull the plug in a tub of water, the surface water starts its revolutions in large spirals; and at the drain point, the revolution of the water becomes much faster. It is similar to a hurricane: in the upper atmosphere, the speed of the wind is slower; and at the point where the hurricane touches the earth, the speed of rotation becomes much greater.

In a spiral, revolution and rotation are intertwined. At the center of the spiral, the revolution is fast and the rotation is small, whereas at the beginning of the spiral, the revolution is slower and the rotation is larger.

Thus, when we speak of a "higher vibrational level," it means that the revolution of light is closer to the apex or center, where light revolves the fastest. If the essence of you, or Source, is at the center, centering and balancing the spiraling light, then as you approach the center the rate of revolution is greatest. That is why the pendulum revolves rapidly when you approach the I AM essence.

The Place of Stillness in Meditation

When we speak about spiraling lights in creation, there are *actually two* spiraling lights: the red/male light, and the blue/female light. These lights are moving in opposite directions, so the spirals of light move through each other.

The nature of creation is that there is an opposition of motion, with lights spiraling from the opposite direction. The result is that at the very center where the two lights meet—the apex of the spirals—there is the greatest motion in opposition. This results in a motion-in-equilibrium, which in turn results in a place of stillness.

This is the point of stillness we encounter when we meditate or bring ourselves into the I AM essence state. This is the place of greatest balance/harmony. This is the location of the fulcrum for the two opposing spirals of light.

A similar phenomenon of stillness also occurs in hurricanes with spiraling winds. We often hear it said that "in the eye of the hurricane, there is stillness." Winds are swirling rapidly, but in the eye of the storm—where one would expect the greatest wind-speeds—there is stillness. This is because two forces of motion are occurring that are opposed, resulting in a state of motion-in-equilibrium.

Stillness in All Matter at Its Core

Since all matter is built in the same way as spirals of light, then all matter has a point of stillness at its essential core—a point of balance and harmony. In fact, every human being has his/her own point of stillness of harmony and balance. We reach that point in deep meditation, or when we go to the I AM level. At that level, there is stillness and balance/harmony.

Walter Russell said, "God is at the center of the fulcrum, balancing everything." He further stated: "And then God said to me, 'Behold thou the unity of all things in Light of Me, and the seeming separateness of all things in the two lights of my divided thinking. See thou that I, the Undivided, Unchanging One, am within all divided things, centering them, and I am without all changing things, controlling them.'"

The Spiral in Breath

Let's examine some very interesting aspects of breath by means of the pendulum. If you test yourself or another person with a pendulum during the *inhalation* of breath, you'll find that the pendulum rotates clockwise, and that during the *exhalation* of breath, the pendulum rotates counterclockwise. Remember from the discussion of spiraling energies that the clockwise rotation is creation-generation, and that the counterclockwise rotation is breakdown-radiation. Thus, breath appears to be related to this cycle.

There is another point that yogis describe, which we can observe when we pay attention to breath: in the middle of the cycle between the inhalation and the exhalation is the momentary point of stillness.

The Spiral in Our Structure

The spiral is in the very essence of our physical structure. The spiral within our structure is that of the golden mean spiral. The Fibonacci sequence creates a spiral that approximates the Golden Mean spiral, so our structure often is described as being related to the Fibonacci sequence or spiral.

The figures below, briefly introduced in Figures 1 and 2, show the typical Fibonacci spiral; the other figure is the Golden Mean spiral. In mathematics, the *Fibonacci numbers* form a sequence—called the Fibonacci sequence—such that each number is the sum of the two preceding ones, starting from 0 and 1. A sequence would be:

$$0,1,2,3,5,8,13,21,34, \ldots$$

A tiling of squares whose sides are the successive Fibonacci numbers would result in the following figure.

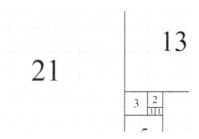

FIGURE 34. Tiling of successive Fibonacci numbers by creating squares of the numbers.

Connecting opposite corners of the squares in the Fibonacci tiling creates a spiral that approximates the Golden Mean spiral, as seen in the figure below.

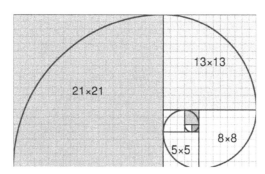

FIGURE 35. The Fibonacci spiral created from the tiling of the squares from the Fibonacci sequence of numbers.

As the numbers of the Fibonacci sequence increase, the ratio of the numbers approximates the Golden Mean of 1.618. I am describing the Fibonacci sequence because most of the literature discusses the structures in nature in terms of the Fibonacci sequence. The sequence can be drawn as shown in the figure above, or as a vortex in a cone shape. The Golden Mean spiral, drawn as a cone, was shown as Figure 31 in the previous chapter.

In the article, "Fibonacci Series, Golden Proportions, and the Human Biology" by Dharam Persaud and James P. O'Leary, one sees the obvious connection between human biology and the Golden Ratio/Fibonacci spirals. A few of the pictures from the article are reproduced below.

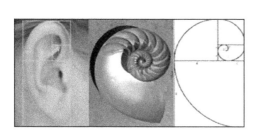 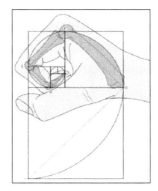

FIGURE 36. Phi in Ear and Hand.

Drunvalo Melchizedek, in his book, *The Ancient Secret of the Flower of Life*, Volume 1, shows the phi ratios in Doryphoros the Spear-Bearer sculpture and the calculation of the phi ratio for the hand (see figure below).

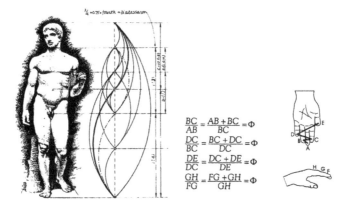

$$\frac{BC}{AB} = \frac{AB+BC}{BC} = \Phi$$

$$\frac{DC}{BC} = \frac{BC+DC}{DC} = \Phi$$

$$\frac{DE}{DC} = \frac{DC+DE}{DE} = \Phi$$

$$\frac{GH}{FG} = \frac{FG+GH}{GH} = \Phi$$

FIGURE 37. Phi ratio in the human body and in the hand.

Summary

The vortex is the basis of creation, as discussed in *The Seeker and The Teacher of Light*. The vortex motion is actually a spiral with the geometry of the phi ratio. That spiral motion becomes more rapid as it approaches the apex of the cone of the vortex. The rate of revolution of the spiral is likely the origin of our vibrational level. The pendulum moves clockwise (the direction of centripetal winding of light for creation) when the vibrational level is measured, and the rate increases as our vibration level moves up.

The spiral shape of the Golden Mean ratio curve is basic to our structure, and is seen throughout nature. The vortex motion in the phi spiral has BG3 as well as 3-6-9, and is likely their origin. Our breath is also linked to the spiral. Breathing in brings the clockwise energies for creation, and breathing out expels the energies and products of the breakdown of matter.

3-6-9 and GANS

The subtle energies of 3-6-9 are associated with the torus structure and the nature of creation. Part V describes the detection and measurement of these energies. GANS (Gas in Nano State), which is fundamental to the nature of matter and plasma fields, can be considered a key element of matter. GANS possesses 3-6-9 subtle energies, and creates fields containing both 3-6-9 and BG3 subtle energies. The details of these discoveries are described in this section.

Measuring 3-6-9 and BG3 Levels of Creation, Torus Symbols, and Creating Powerful Arrays

The 3-6-9 Energies and the Symbols of Creation

The meaning and importance of 3-6-9 was explored in detail in *The Seeker and The Teacher of Light*. In this book, some of these concepts were re-introduced in Chapter 3. In this chapter, we will go into more detail and discuss the measurement of 3-6-9.

In *The Seeker and The Teacher of Light*, we discovered that the numbers 3-6-9 represent the torus structure and the creation principle. Throughout history, there have been symbols related to creation. We discovered that these symbols have subtle energies associated with the torus. The symbols we are most familiar with are the symbols of aum, yin yang, and the bagua. We have labeled these subtle energies as *3-6-9*.

Understanding the symbols, their construction, and their use will help to give evidence on the nature of creation and give us a greater understanding as to who we are.

With the discovery of the 3-6-9 pendulum, we found that we could be in resonance with the subtle energies represented by the 3-6-9 symbols. The symbols were also in resonance with BG3, the BioGeometry quality of harmony and the center. However, we learned that BG3 and 3-6-9 energy qualities sometimes interfered with each other when they were placed on top of or sometimes next to each other, canceling the energy qualities of both. We also learned that *the thought/intention of harmony* eliminates this interference. The energies of the yin yang symbol for harmony also cancel the interfering qualities of BG3 and 3-6-9.

As a reminder, the 3-6-9 energy qualities are found in the symbols below. Since the universe looks at *units* to represent numbers, units (as lines) have 3-6-9 energy qualities; and the numbers associated with Marko Rodin's Vortex Math, which translates to the torus structure, also show 3-6-9 energies.

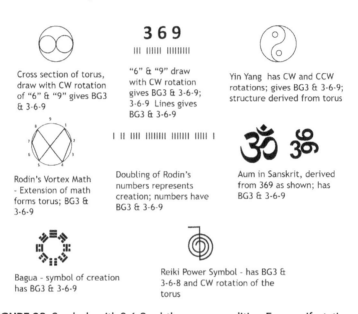

Cross section of torus, draw with CW rotation of "6" & "9" gives BG3 & 3-6-9

3 6 9

"6" & "9" draw with CW rotation gives BG3 & 3-6-9; 3-6-9 Lines gives BG3 & 3-6-9

Yin Yang has CW and CCW rotations; gives BG3 & 3-6-9; structure derived from torus

Rodin's Vortex Math - Extension of math forms torus; BG3 & 3-6-9

Doubling of Rodin's numbers represents creation; numbers have BG3 & 3-6-9

Aum in Sanskrit, derived from 369 as shown; has BG3 & 3-6-9

Bagua - symbol of creation has BG3 & 3-6-9

Reiki Power Symbol - has BG3 & 3-6-8 and CW rotation of the torus

FIGURE 38. Symbols with 3-6-9 subtle energy qualities. For manifestation of clockwise (CW) rotations for BG3, the direct torus-related symbols need to be drawn in a CW manner for the "6" and "9" aspects of the torus and the 3-6-9 numbers. For Yin Yang, the center line must be traced to activate the symbol. For Rodin's symbol, the line connecting the numbers must be traced for activation. The Reiki Power symbol is normally activated, since it is drawn from the center in a clockwise motion; when drawn counterclockwise, it is not activated and shows negative BG3 (counterclockwise rotation of BG3 pendulum).

However, merely having a picture of the symbols did not always produce the 3-6-9 qualities. In Chapter 6: "The Source of Spirals," we learned that the clockwise motion is the creative, generative direction, and that the counterclockwise rotation is the radiative, disintegrative direction, which is not conducive for health/harmony.

The same rules apply in drawing the torus and writing the numbers 3-6-9. Clockwise directions in drawing the circles in the torus, or the direction of rotation in drawing the numbers 6 and 9, result in clockwise BG3; and the counterclockwise direction of drawing the circles or the 6 and 9 results in a counterclockwise rotation for BG3 (i.e., negative resonance effect). In order for the resonant energy qualities of 3-6-9 to show up in the figures of the torus, the numbers of 3-6-9, and the yin yang, the circles or numbers need to be traced in the correct direction. Once the numbers have been traced, then the energy is manifested and is permanent. The images can even be photocopied, and the energy will remain with the photocopy.

What this means for you is that you can use the numbers or symbols to create harmony and balance within yourself and your environment, since the symbols and numbers have BG3 and 3-6-9 qualities. The numbers and symbols only need to be photocopied and carried with you; or you can make them easily at any time by drawing them on paper. With a harder surface, they can be written using a carbide pen.

Who We Are and the Nature of Creation

This finding suggests that there are key lessons to be learned about who we are and the nature of creation/reality:

1. **We are able to sense energy qualities (BG3 and 3-6-9) through the focusing aid of a neutral pendulum.** These are invisible energies; a typical Western scientist would say, "Impossible!" but you can see for yourself by trying it out, based on what you have read so far.

2. **Thought/intention is powerful and measurable.** For example, the interference of 3-6-9 symbols and BG3 tools that have nullified the ability to detect these energies can be reversed by merely a *thought* of bringing in harmony. You can also bring in harmony by means of the yin yang symbol. This means that your thoughts can change reality.

   Example 1: In any situation where there might be disharmony, you can invite in harmony. When you are working on yourself towards healing and some cells are causing you discomfort (e.g., pain), you can *invite in harmony with the disharmony.* Your body then realizes that there is disharmony, and will bring back the natural harmony.

Example 2: Another example is the meditations for world peace that took place using Transcendental Meditation. Scientists theoretically predicted that the square root of just 1 percent of the U.S. population (about 1,725 people) meditating together would create a nationwide positive effect. The spiritual teacher who brought Transcendental Meditation to the West, Maharishi Mahesh Yogi, showed that this form of meditation was linked to decreased homicides and violent crime in the U.S. from January 2007 through 2010. During this three-year study period, over 1,725 participants gathered to practice group Transcendental Meditation at Maharishi University of Management in Fairfield, Iowa. At the end of the study, researchers found an 18.5% drop in violent crime nationwide. Statistical and independent analyses showed that the rising trend of U.S. homicides that had occurred between 2001–2006 was reversed during the 2007–2010 study period. (Other variables were ruled out, such as the economy, demographics, and law enforcement.)

3. **Symbols are not just abstract symbols. They have energies,** which can be detected and measured (BG3 and 3-6-9).

4. **Drawing the symbols produces the *effect* of the symbols.** A torus or a 3-6-9 set of numbers can have a positive resonance or a negative resonance, depending on how you draw the image/symbol, even though the images look the same. What you are drawing affects the reality of what you draw.

5. **The intention behind your drawings or words, when put in print, imbeds the meanings in the print.** These drawings or printed writings can be photocopied, and the power will remain with the words and symbols even in the photocopied version. Words and symbols have the power of thoughts behind them.

6. **The spiraling direction of the rotation of energies indicating that the creative/generative cycle goes clockwise is further validated,** as is the case with the breakdown/radiative cycle going counterclockwise.

7. **The BG3 and 3-6-9 directions of rotation of pendulums are further explained** in light of the concept of centripetal clockwise motion being the direction of the creation/generation part of the cycle of creation, and centrifugal counterclockwise rotation being the direction of the radiation/breakdown part of the cycle.

Measuring the 3-6-9 and BG3 Levels of Symbols

For those readers who are interested in the levels of BG3 and 3-6-9 as measured with the BG3 and 3-6-9 pendulums, I have created a table of the symbols and the respective values of the two energy qualities. The values in the tables represent numbers I obtained using the BG3 ruler described below. There can be some variability in values, depending on the consciousness of the person using the pendulum at any given time. However, among BioGeometry practitioners, this variability is not typically more than a few relative units on the ruler scale.

The tool used to obtain these measurements is called a *BG3 ruler*. This special ruler is used because subtle energies are hindered by the 90-degree "L" shape. If you have a series of these "L" shapes, the greater the energy, the further you can go before the "L" has stopped the energy. You would use a BG16 pendulum for BG3 measurements, and a 3-6-9 pendulum for 3-6-9 energies. The two energies typically give identical results, but not always. Arrays will nullify BG3 tools.

Below is a schematic to show the principles of a BG3 ruler used to measure levels of BG3 or 3-6-9. (It is not to the scale used in Bio-Geometry, since that belongs to BioGeometry and is available only in their Foundation Course.) You can see an actual image of the BG3 ruler on this page of the BioGeometry website (see image for BG3 Scale with Sphere): *https://biogeometryeurope.com/product/bg3-scale-with-sphere-attachment/*

The units on the ruler are relative and based on its construction. BioGeometry practitioners use the BioGeometry BG3 ruler, so the relative scale is standardized to that ruler. The numbers used in this book are based on the official BG3 ruler.

The principle of the ruler is that a 90-degree angle has blocking abilities with respect to BG3 and 3-6-9. The greater the amount of BG3 and 3-6-9, the higher the number on the ruler where you can still measure BG3.

The object you are measuring is placed in the "Witness circle"—a circle in which the object to be measured is placed. This circle has an opening, thus allowing subtle energies to move in the direction of the opening. Detection of BG3 subtle energy is determined by a clockwise rotation of the BG16 pendulum. For example, if you have a relative BG3 level of 500 for a glass of water that has been placed in the Witness circle, then your BG16 pendulum will detect BG3 at the apex of the 90-degree point up through 500—that is, at 100, 200, 300, 400, and 500. However, no BG3 would be detected at positions 600 and beyond.

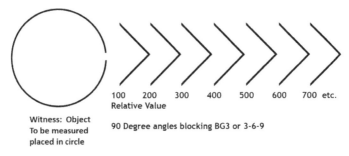

100   200   300   400   500   600   700 etc.
Relative Value

Witness: Object
To be measured   90 Degree angles blocking BG3 or 3-6-9
placed in circle

FIGURE 39. Schematic (not to scale) to show principle of BG3 ruler (also works for obtaining relative values of 3-6-9). Object to be tested is placed in the Witness circle, and a BG3 pendulum (BG16 or IKUP pendulum) is used to measure the relative level of BG3 or 3-6-9. Readings are made at each 90-degree line until no more BG3 is detected. (The picture is not standardized to the size used in BioGeometry. That is available in the BioGeometry Foundation Course: *https://www.biogeometry.ca/courses-events*.)

Below is a table of relative values for BG3 and 3-6-9 for the various symbols from the BG3 ruler. Values will vary slightly from day to day and person to person, but will be within the ballpark of the values above. I have added a column with a BG3 tool, the L90 BG3 emitter. The table shows that when L90 and a 3-6-9 symbol are put together, both BG3 and 3-6-9 subtle energy qualities are nullified, resulting in no detection of either energy qualities. However, putting in a yin yang symbol, which represents harmonization, causes the energies of BG3 and 3-6-9 to return. As described in *The Seeker and The Teacher of Light,* having the conscious intention of harmonizing the BG3 tool and the 3-6-9 symbol will also eliminate the nullifying effect of the combination of the BG3 tool and the 3-6-9 symbol.

Interestingly, aum has the highest level of BG3 for the symbols. My friend who does channeling says this is because aum is the oldest symbol, and its energy has gradually been enhanced with use over time. Aum still represents the creation symbol, as do the other symbols.

Symbol	BG3 Value	3-6-9 Value	BG3 Value Symbol + L90*	BG3 Value Symbol + L90* + ☯
ॐ	1000	1000	0	1200
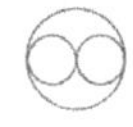 ∞	100	100	0	400
3 6 9	100	100	0	400
 ☯	100	100	100	400
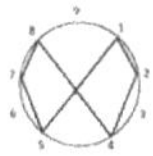 ⊗	100	100	0	400
 ✳	100	100	0	400
 ⊚	100	100	0	400

* L90 Is a BioGeometry tool which emits BG3

FIGURE 40. The table illustrates the relative BG3 and 3-6-9 values of the various symbols, as well as the nullifying effect of the BG3 and 3-6-9 values when the two types of symbols are put together. The last column shows the return of the subtle energy resonant values when a harmonizing yin yang symbol is added.

The 3-6-9 and BG3 Symbols and the Power of Their Arrays

We have learned that the 3-6-9 pattern of creation found in key creation symbols is in resonance with BG3 and 3-6-9 subtle energy qualities, showing significant levels of BG3 and 3-6-9. *This made me wonder whether those energies can be magnified to higher levels.* Some applications might be as an aid for healing, and for making food, supplements, and water more beneficial for ourselves.

An array is a geometric arrangement of the symbols. A square array would be 4 symbols at the corners of a square shape. A special 3-6-9 array is described in this chapter.

The Power of the 3-6-9 Array Geometry

Understanding the power of 3-6-9, I decided to take the aum symbol, which has a BG3 value of 900-1000, to see if it is possible to increase the level of BG3 and 3-6-9. As shown below, I placed the aum symbol in a 3-6-9 array. There would be 3 aums in the center, 6 aums in a further larger circle, and 9 aums in the outer circle. This array now produced the very large BG3 and 3-6-9 levels of 6,400, as compared to a single aum level of 900–1000.

FIGURE 41. Aum placed in a 3-6-9 array. The 3 aum symbols are in the center, surrounded by 6 aums, and finally 9 aums in the outer circle. This array produces BG3 and 3-6-9 of 6,400.

However, even though aum is used in this example, the other 3-6-9 symbols should also give increased levels of BG3 and 3-6-9 in this configuration. It should be further noted that if you print the aum and their arrays, and then place a BG3 tool (such as a cube for harmonizing a space) on top of the printed aum or their array, both BG3 and 3-6-9 will no longer be detected—that is, the BG3 and 3-6-9 subtle energies have been nullified. The interference is harmonized (eliminated) either with the intention of harmonizing or with the use of a yin yang symbol.

The Further Increased Power of a 3-6-9 Fractal of the Array

The next step was to miniaturize the 3-6-9 array with computer graphics. I then put the miniature array into another 3-6-9 array. Thus, we have an array of an array. This produced the huge BG3 and 3-6-9 subtle energy levels of 17,500 on the BG3 ruler.

FIGURE 42. The 3-6-9 array of aum symbols was further reduced in size, and each array was placed in another 3-6-9 array, creating a fractal of the 3-6-9 arrays. This resulted in a BG3 and 3-6-9 level of 17,500.

What this means is that it is thus possible to produce as large a level of BG3 and 3-6-9 as desired for any particular application.

I then put a vial of tap water on top of the array to see how the array would activate the water. The result of that experiment is shown below. Depending on the duration of the wait time, the array brought the water to a level of 7,100 in 96 hours. This is a very high level of activation, or structuring, of water. The best spring water is very seldom much more than around 1400. The control was tap water not placed on any array.

3-6-9 Aum Array of Array - Effect on Water

Time (Hr)	Control BG3	Control 3 - 6 - 9	Array* BG3	Array* 3 - 6 - 9
0	200	200	200	200
0.2	200	200	600	600
5	200	200	1500	1500
13	200	200	2000	2100
30	200	200	3400	3400
48	300	300	4500	4500
72	300	300	6400	6400
96	300	300	7100	7100

*369 Aum Array of Array

FIGURE 43. The kinetics of water activation/structuring, as evidenced by the increase in BG3 levels when water was placed on top of a 3-6-9 array of a 3-6-9 aum array (fractal pattern in 3-6-9 geometry in above figure). In 96 hours, the BG3 of the water reached 7100. A control water was run at the same time, which did not show increase in BG3 and 3-6-9 values.

An Even Further Increase in Power of the Arrays with Spiritual 12

Knowing that aums and other 3-6-9 symbols interfere with BG3 tools when they are used together without harmonization, my friend Greg mentioned that I should see if the use of 12 would create further harmonization. There are spiritual connotations with 12—e.g., the 12 disciples of Christ.

I created the 3-6-9 array of aums, and put a further 12 aums surrounding the 3-6-9 array. This, shown below, resulted in a BG3 and 3-6-9 level of 13,500. This array did not cause interference with BG3 tools. With the 12, harmonization was built into the array.

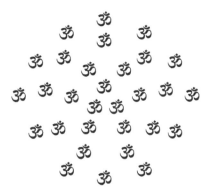

FIGURE 44. Spiritual 12 aums surrounding the 3-6-9 array of aums resulted in a BG3 and 3-6-9 level of 13,500, as well as harmonization of the nullifying effect of 3-6-9 symbols on BG3 tools.

Significantly Increased Power of a 3-6-9 Fractal of the Array with Added Spiritual 12

When you take the 3-6-9 array of aums with the additional surrounding 12 spiritual aums, miniaturize them, and then arrange those arrays in a 3-6-9 pattern, then further enclose the resulting array with 12 aums, you get the fractal pattern below. The resulting fractal now shows BG3 and 3-6-9 levels of 135,000. The numbers are so very high that it is difficult to say definitively that 135,000 is the actual number, but I can say that the number is very large. The level is way beyond any conventional BG3 rulers; special rulers were invented to make this measurement.

FIGURE 45. Fractal 3-6-9 Array of 3-6-9 aum array with 12 surrounding aums at each level. This produces the highest BG3 and 3-6-9 levels in the range of 135,000, completely off the scale of conventional BG3 rulers.

I next decided to use this fractal array to determine the kinetics of water activation. The results are shown below.

Fractal 3-6-9 Aum + Spiritual 12 Array Activation of Water

Time (Hr)	Control BG3	Control 3-6-9 Energy	Test BG3	Test 3-6-9 Energy
0	100	100	100	100
0.2	100	100	600	600
0.5	100	100	1000	1000
2	100	100	1900	1900
4.5	100	100	2700	2700
7.5	100	100	3500	3500
14	100	100	4300	4300
23	100	100	7400	7400
34	100	100	8200	8200
39	100	100	8400	8400
48	100	100	8700	8700

FIGURE 46. Activation/structuring of water using the Fractal 3-6-9 Aum Array with Spiritual 12 aums with each array, as shown in Figure 45.

Arrays are interesting ways to generate very high levels of BG3 and 3-6-9. The effect of very high levels of BG3 and 3-6-9 can be a subject for future investigations by scientists and others who are curious about this very compelling field.

Summary

The purpose of this chapter was to demonstrate the following points:

- We have the ability to detect and to quantitate the level of BG3 and 3-6-9 of the creation torus symbols using BG16 and 3-6-9 pendulums.

  The power of the resonance of the symbols can be demonstrated and photocopied, based on how the symbols are drawn (e.g., a cross-section of a torus drawn as a larger circle and 2 smaller circles). If the symbols are drawn as just 3 circles, then there is no BG3 or 3-6-9. If 2 smaller circles are drawn in a clockwise direction, there will be BG3 with clockwise rotation of BG16 pendulum. If the symbols are drawn in a counterclockwise manner, the pendulum will rotate counterclockwise. Symbols retain the energy given to them by how they are drawn; that energy can be photocopied.

- BG3 tools and 3-6-9 symbols, when put together, can nullify their BG3 and 3-6-9 effects. Use of the yin yang symbol or the conscious intention to harmonize the tools and symbols will result in harmonization.

- Geometric arrays can be used to increase levels of BG3 and 3-6-9.

- Since Dr. Ibrahim Karim has demonstrated that increased levels of BG3 result in activating/structuring water, as evidenced by his work with Masaru Emoto's water freezing/crystallization experiments, we have demonstrated that water can be significantly activated using the creation torus symbols.

CHAPTER 9

The 3-6-9 Energies of GANS, Ormus, and Their Fields

Everything in the universe is made from magnetic fields and their interactions, from the atom to the body to everything in the universe. The field is composed of spiraling forces that move toward a center through gravitational forces and outward through magnetic forces. Matter is formed in the process.

For many of you, the world of GANS (Gas in the Nano State) and Ormus will be a new area. I hope that by the end of this chapter, you will find knowing about this area and its applications to be well worth your while. You will learn wonderful lessons about yourself, your nature, and the many applications that you can use for your benefit. In addition, what you will learn from the study of GANS/Ormus will also carry over to important applications in the world of 3-6-9 energies.

Why do I think it's important to tell you about the world of GANS and Ormus?

- For those scientists/experimenters in the new brave world of GANS and Ormus, I will introduce to you to a new tool for measuring the energies that you are creating. Right now, you are only guessing at the location of the fields and their strength. The tool will allow you to detect and actually measure the fields created by these materials.

- For those who want to create fields with greater levels of BG3 or 3-6-9, I will describe how this can be done.

- For those who still doubt their true I AM nature, I will show you how you can create fields that you can measure, and *how to move or copy them by thought/intention.* It is easy to do; everyone can do it. You can detect the movement or copying and can prove that the energy has moved or copied to the new location. This helps demonstrate that *you are your thoughts.*

- The elementary GANS materials have characteristics of the spiraling energies described by Walter Russell. As a result, they are further evidence of the nature of creation. The fact that the GANS energies/fields can be measured by the 3-6-9 pendulum tools is further evidence of the toroidal nature of the energies.

In this chapter, you will learn about the nature of GANS and about the fields of BG3 and 3-6-9 that they can create. In the next chapter, you will be invited to conduct some experiments that show how you can move and copy fields that you make with conscious intent. By doing this, you will have a clearer idea as to who you are.

3-6-9 Energies and Monatomic Materials–GANS, Ormus

In this section, I will describe the nature of GANS and Ormus, and then discuss the fields they make. The main interest, from this book's perspective, is the utility of the fields to generate BG3 and 3-6-9, as well as our ability to interact and direct the fields produced by means of thought/intent/consciousness. From studying GANS, you will again see the spiraling nature of creation described by Walter Russell and others. You will also learn that GANS has tremendous potential to change the way we work with matter and energy.

Nano State, GANS, and Plasma States of Matter

To understand the new terms "Nano," "GANS," and "Plasma States," I will describe how these states of matter are made. If I just gave you definitions, they would be too abstract.

A new science and way of looking at how the universe is put together has been created by Mehran Keshe, a scientist from Iran. In this chapter,

I will be using graphics and terminology taken from Mehran Keshe's website (*https://keshe.foundation/education*). Although the terminologies are not necessarily in common use in the fields of physics and chemistry, they are descriptive of what Keshe observed, and they coincide with my own observations in researching the subject.

Let's discuss the figure below and the terms used.

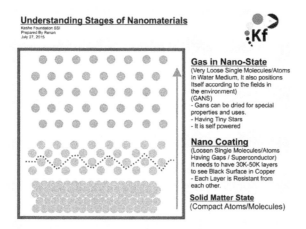

FIGURE 47. Understanding the stages of nano materials. The figure summarizes the transition from solid matter to the monoatomic nano layer and the separation to the freed-up monatomic state called GANS. Both nano state and GANS show 3-6-9 subtle energy qualities.

The Nano State

In the above figure, let's assume that the bottom layer is a copper wire (solid-matter state). If you subject the copper wire to a flame (e.g., propane torch), a black coating occurs on the copper wire. Copper does not burn and it does not melt at the lower propane-flame temperature. What has happened is that the surface layers of copper have separated from the solid state of copper. The blackened copper is called the *nano state* of copper and has different properties. Keshe describes the super-conductivity of copper as being due to the nano layer. The copper atoms are now single atoms with gaps between them, which allow electrons from a current to move through with little resistance.

From my observations, the nano layer now has 3-6-9 subtle-energy qualities, which is detected by the 3-6-9 pendulum. Since the atoms are single atoms, the copper can be considered "monatomic." Upon formation of the nano layer, this layer has a negative polarity. The nano layer thus remains associated with the physical, more dense copper matter with a more partial positive polarity. The nano layers are not independent but interact with the physical matter, with each other, and with the environment through magnetic and gravitational fields (Mag-Grav Fields).

Another way to form the nano layer of copper is to place a copper plate or wire in a caustic environment (vapor of sodium hydroxide, aka lye). As with the flame, the surface layers of copper are separated and form a monatomic state or nano layer above the physical matter (the copper). This is shown in the figure below.

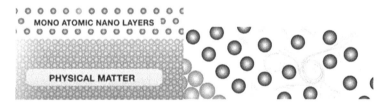

FIGURE 48. Formation of monatomic nano layers with use of a caustic environment. The close-up portion of the figure shows the gaps and MagGrav Fields—the center of energy and information exchange.

The term *nano* refers to one billionth of a meter (a sheet of paper is 100,000 nanometers thick). In Keshe's terminology, *nano* refers to entities at the nanoscale, such as atoms, which are independent structures. Other people have used the term *monatomic*. The atoms are independent structures and behave like round magnets on a surface.

GANS (Gas in the Nano State)

Certain conditions will free the monatomic nano layers from the physical matter with which they are connected. An example would be freeing the nano copper layer by treating the nano-coated copper plate with a salt-water environment. The freed monatomic copper is now in the salt water and is called *GANS* (for *Gas in the Nano State*). The GANS will continuously be giving and receiving fields (MagGrav Fields). The term "Gas" in GANS does not necessarily mean a gas in the standard way we visualize it. Gas in the GANS means *a free-standing monatomic state*. The GANS can be detected with the 3-6-9 pendulum.

MagGrav Fields

Magnetic fields are constantly interacting. If you have four round magnets on a small surface and you move one magnet around, the other magnets will also move and adjust their location, based on the magnet you are moving. If you consider a bar magnet, it has two poles—one attracts and the other repels. A "field" really implies both a *magnetic field* and a *gravitational field*. The magnetic pole repels, and the gravitational pole attracts. Keshe has coined the term *MagGrav Fields* to describe these fields. The strength of each magnet will dictate the gap between the magnets as you move the magnets around on a surface.

Consider two magnetic spherical systems. The field exiting the top of one magnet interacts with the bottom of the other magnet. This is how the magnets adjust to keep a gap between themselves. Keshe's drawing showing this phenomenon appears below.

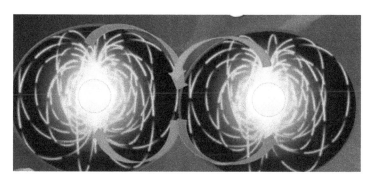

FIGURE 49. Interaction of the fields of two magnets, or two GANS units, or monatomic units, or units in the nano layer. The gravitational/magnetic interactions dictate the gap or distance between the magnets or units comprising the MagGrav Field.

In any field of a magnet, there is both the Magnetic flow out and away from the magnet (repulsion), and the Gravitational flow toward the magnet (attraction). When talking about magnets, we refer to the "north pole" and "south pole" of the magnet. However, a better description of this phenomenon is "gravitational" and "magnetic." The fields they produce would be termed *the magnetic field* and *the gravitational field*. This is the principle of the MagGrav Fields. Keshe states that *everything* in this universe is made of fields.

The MagGrav Fields associated with the nano layer and GANS can then capture other plasma magnetic fields of matching strength. The existence and interactions of dynamic and different strengths, speeds, and densities of magnetic fields lead to the formation of matter—atoms, molecules, plants, animals, planets, and galaxies. In other words, the nano and GANS state results in formation of MagGrav Fields, or plasma fields, which in turn can lead to matter formation. These fields have "conditioned" the space so that the space now shows BG3 as well as 3-6-9 subtle energy qualities.

Everything in the universe is made from magnetic fields and their interactions, from the atom to the body to everything in the universe. The field is composed of spiraling forces that move toward a center through gravitational forces and outward through magnetic forces. Matter is formed in the process.

The MagGrav nature of both the nano and GANS state of matter is similar to the Walter Russell vortex nature of matter, which is the basis of the torus concept of matter formation. Russell often describes the gravitational/magnetic nature of the vortex for matter formation.

Since there is so much that might come from understanding the science that Keshe is proposing, and since one person or group cannot do everything, he has tried to give away the fundamentals of his understanding of the universe, and has created the Keshe Foundation Space Ship Institute to teach the fundamentals. His series of about 12 YouTube presentations and his website describe the basics of his beliefs: *https://www.youtube.com/playlist?list=PLpCKWzA-bp9unXm9drxd-DwX82l6QtYqc-* and *https://keshe.foundation/education.*

A Broader View of GANS and Plasma

As Keshe's work shows, our bodies, and life in general, are composed of billions of GANS at various levels of interaction, forming the total plasma of the body. The most abundant types of GANS for the body would be carbon, oxygen, hydrogen, and nitrogen.

Plasma is the invisible spiraling field behind everything, from the vacuum to dense matter. Plasma can be considered the foundation for a single MagGrav unit or a collection of units to form an atom or a proton. On a larger basis, a planet such as Earth can be considered a plasma. The sun, the solar system, the galaxy, and the universe can be considered a plasma. All things are collections of fields, arranging themselves in different states because of different environmental conditions.

Gravity is due to the interactions of magnetic fields. The magnetic fields that make up our body come from the fields of the sun, the galaxy, and the universe. The whole world is made of MagGrav fields. When fields slow down enough, they become matter. The different elements of earth are derived from the interaction of the fields of the sun and fields of Earth. In the lab, conditions have been created to generate the GANS of carbon, hydrogen, oxygen, and nitrogen.

A summary of the various states of matter is described in the useful figure below.

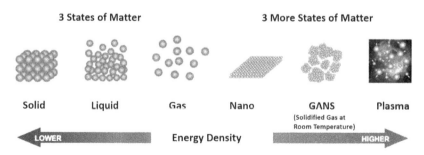

FIGURE 50. The various states of matter and their energy density. The nano state is the fourth state of matter. It is a "CAPTURED" and CONCENTRATED GAS–in the magnetic-gravitational plasmatic state.

The nano layer can be considered the fourth state of matter (solid, liquid, and gas being the three states of matter commonly described). It is both plasma-based and associated with normal condensed matter, because of polarity differences in the nano state as compared to the physical-matter state. Both the nano layer as well as the freed GANS state radiate their own spinning magnetic plasma energy fields (e.g., the sun), while drawing in gravitational plasmatic fields. It has a similarity to the torus in terms of spiraling fields.

The nano state is sometimes described as a "captured and concentrated gas in a solid state, with distinct magnetic and gravitational properties." Keshe states that in the magnetic-gravitational plasmatic state, the nano particle, or GANS, radiates out its own spinning magnetic plasma energy fields, just like the sun, while also drawing in gravitational plasma fields.

Thus, the appearance resembles the torus, with spiraling magnetic fields pushing out and gravitational fields pulling in. The spiraling nature of the energies is reminiscent of the vortex described by Walter Russell, or the basis of matter formation from the torus structure. The centripetal motion inward would be the gravitational field for creation, and the centrifugal motion outward would be the magnetic field for de-creation. In Russell's work, the centripetal clockwise motion for compression of light is described as "gravitational" and sometimes "electrical." The outward centrifugal motion is often described as "radiation" and sometimes as "magnetic."

A container of GANS (or Ormus) would consist of millions of "little suns," all with magnetic and gravitation fields. This would be similar to a bunch of circular magnets with north and south polarity. On a flat surface, the magnets would move around and adjust their position based on the field of the magnets. A similar situation would occur with the monatomic Ormus, or GANS. The individual GANS units would form an array within the water (for liquid GANS), based on their individual magnetic/gravitational properties. As mentioned, Keshe uses the term "MagGrav" to describe the fields that are produced. The interaction of the MagGrav fields between nano plates and other matter is what creates the condition that allows GANS to manifest. For example, a copper plate coated with copper nano particles would manifest CO_2 GANS in the presence of a zinc plate.

The Vibrational Frequency of GANS and Its Interaction with Human Intention

According to Keshe, we now know that GANS is another state of matter at its atomic level. GANS vibrates at a perfectly balanced frequency. It can be used to attract and repel, or to bring another similar field into balance. One might say that it holds a perfectly tuned note (or notes), which enables the body systems to re-tune, or come into focus.

Anything can be converted to a GANS state, in fact. One of the primary GANS for this technology is CO_2/ZnO from the copper- and zinc-plate method. However, other metals and non-metals also can be used to make GANS, such as calcium, food, seawater, soil, etc.

GANS connects with the environment, and receives and radiates plasma fields. It is believed to also interact with human intent, and is capable of learning how electricity in our homes and buildings behaves. There is one device already on the market that enables the plasma fields to start to mimic the behavior of "normal" electricity and deliver plasma energy to devices that we want it to power.

Keshe worked out how to make just the right alkaline and temperature environment to encourage single atoms to migrate and be stable as monoatoms on the surface of a metal. He has shared his techniques for making GANS and nano layers at his Keshe Foundation Spaceship Institute, with the idea that people can figure out innovative ways to use the GANS technology. He talks about the many applications for these materials (agriculture, food, decontamination, power generation, healing).

Ormus

Ormus is the first GANS to be discovered. At the time of discovery, the term "GANS" had not yet been coined. The story of Ormus is interesting. Following is a short synopsis of some of the many stories.

Amazing Stories about Ormus

In the 1970s, David Hudson, a wealthy cotton farmer, found mysterious elements in his volcanic soil that defied normal analysis (i.e., they had no recognizable spectroscopic signature). Electrons were re-arranged in such a way that the elements no longer participated in chemical reactions. Since they lacked bonds holding metal atoms together in the solid form, they were called "monatomic." They are often referred to as the "M-state" of these elements.

The original name given to these materials was Orbitally Re-arranged Monatomic Elements, or ORMEs. However, more recent work indicates that some may be diatomic (having two atoms rather than one), and also that very loose bonds can exist with certain other elements, particularly alkali metals. The generic name has been changed from ORMEs to Ormus.

I should reiterate that the science of GANS and Ormus is in its infancy, since analytical tools to study these materials are sparse. Academic researchers have typically stayed away from studying GANS and Ormus because typical analytical tools are not useful for examining these materials. The 3-6-9 pendulum is the first tool that can detect GANS and their fields. Academic scientists who have not seriously studied radiesthesia or BioGeometry would not risk their reputation talking about using a pendulum for analysis. Academic scientists have typically ignored the study of GANS and Ormus, since tools are lacking for their study.

David Hudson and other researchers have found Ormus in:

- *Most soils tested for Ormus*, especially volcanic soil. (Remember that heat/fire helps to create/liberate GANS.)
- *Most waters,* especially sea waters, have high levels of Ormus (such as from the Dead Sea).
- *Air*. One researcher has even found it in the air. In the monatomic form, it would behave like a rare gas.
- *Some plants*, especially if grown on volcanic soils (especially Aloe Vera).
- *Calf- and pig-brain tissue*. When these were analyzed, they were found to comprise a remarkable 5% of the dry-matter content.

Some extraordinary phenomena relative to Ormus have been observed but not understood:

- On drying aqueous Ormus Rhodium on filter paper in sunlight, it disappears with an intense flash of light, but no sound or shock wave.

- When working with Ormus Iridium and subjecting it to repeated heating to red heat and then cooling, its weight oscillates violently with each cycle of annealing—i.e., it can lose and then regain *all* of its weight.

- Several investigators have observed another strange phenomenon: drops of liquid containing concentrated Ormus forming on the *outside* of closed containers—especially under the influence of magnetic fields.

- Also unexplained is that certain Ormus preparations in aqueous solution acquire a static-electric charge.

- Ormus can be discharged with an audible spark, after which the charge will re-form, and the Ormus can again be discharged an indefinite number of times.

- Ormus atoms can pass through the walls of containers by quantum tunneling. This refers to some of the Ormus appearing outside of the container—a seemingly impossible barrier. It may be associated with the dual nature of matter, sometimes behaving as a wave rather than a particle.

The stories of the biological effects of Ormus are amazing. A few examples:

- A cat who lost a part of its tail growing back the tail.

- Oranges growing to the size of grapefruits.

- Mortality and infection rates of broiler chickens being reduced by 85%.

See the pictures below:

Ordinary orange   ordinary grapefruit   2 year ORMUS orange   4 year ORMUS orange

FIGURE 51. Pictures from the Internet showing anomalous effects of Ormus on growth of a cat's tail that has been cut off, and the tremendous growth of fruits.

Ormus is one of the easiest GANS to make. Its usefulness for creating BG3 and 3-6-9 fields is a good reason for my describing a simple way to make this form of GANS. For the more complex types of GANS and types of conditions for manifesting them, go to Keshe Foundation Spaceship Institute:

https://www.youtube.com/playlist?list=PLpCKWzA-bp9unXm9drxd DwX82l6QtYqc- and *https://keshe.foundation/education*

Sodium Hydroxide (Lye) Method of Making Ormus

Barry Carter has devoted the last few decades to understanding and getting out the word about Ormus. His website, *http://www.subtleenergies.com*, describes an easy way of making Ormus. The procedure involves merely raising the pH of a solution of sea salt to between 10.6 and 10.78, using a solution of sodium hydroxide (lye). The resulting precipitate—mainly magnesium and calcium hydroxides—carries down the Ormus elements with it. It only remains to wash out the bulk of the salt.

Sodium Carbonate Method of Making Ormus

The use of sodium carbonate is the simplest way to raise the pH, since it avoids working with sodium hydroxide (lye). Below is the procedure I use:

1. Take 50 grams Dead Sea salt (or other appropriate sea salt of good quality).

2. Dissolve in 500 ml water.

3. Take 30 grams pure Soda Ash (sodium carbonate).

4. Dissolve in 500 ml water.

5. Mix the two solutions to obtain Ormus. Stir to mix.

6. Allow Ormus to settle for a day.

7. Then draw off the clear-water layer and discard.

8. If the Ormus is for consumption, add more water and allow to settle overnight, then draw off clear-water layer.

9. Repeat as many times as desired to get a less "salty" taste for the Ormus.

Note that if your research is for the purpose of studying the fields of Ormus and for applications of those fields, then repeated washings are not needed.

For more concentrated Ormus or for drying of Ormus, be aware that Ormus will continue settling, and you would then continue drawing off the water layer. The concentrated Ormus can then be allowed to dry.

Spreading Ormus slurry on a larger surface helps with the drying process.

Instilling BG3 into Vials of Ormus and Water by Instilling Love/Harmony

Some researchers instill Love into the mixture while making Ormus. If that is done, then the resulting Ormus will have BG3 in addition to 3-6-9. If the Ormus is made as just a chemical mixture of sea salt and alkali, then the resulting Ormus will have just 3-6-9. However, if you then take the vial of Ormus with just 3-6-9 and give it Love/Harmony, the vial will become harmonized and have BG3.

This again demonstrates the ability inherent in all of us to make changes with simple intention. We make changes in physical matter with our thoughts.

Since the fields created by GANS or Ormus contain both BG3 and 3-6-9 energies, will water that has been activated by a field contain BG3? The answer is yes. A vial of tap water (BG3 of 300) was placed in a field created with 4 bottles of Ormus. Within 1 hour, the water had a BG3 level of 3000, a very high level.

My home has a reverse-osmosis system for purifying water. I now tape 4 small vials of Ormus around the reservoir containing the purified water. My drinking and cooking water now routinely has a level of BG3 of 3200 at all times. The smaller vials take a little longer for water activation. When the water was tested after the small vials of Ormus were left on the reservoir overnight, the water was at 3200.

Brown's Gas

There is another new material that deserves to be described in this section on GANS. People working in the field of GANS have sometimes called Brown's gas "monatomic hydrogen," and may not even acknowledge that Brown's gas is a GANS material. For that reason, I decided to investigate this material.

My first step in the investigation was to look at a schematic of how to make Brown's gas. By now, my definition of GANS and Ormus had become "the ability to possess 3-6-9 energies."

I have also studied the field of radionics, which teaches that you do not have to make a "nuts-and-bolts" machine in the world of subtle energies. It turns out that just a *schematic* of the nuts-and-bolts machine is sufficient to utilize a radionics machine. These machines were used for healing purposes. My testing showed that the output of the Brown's-gas machine would have 3-6-9 energies. This led me to build a Brown's-gas machine.

The physical and chemical properties of Brown's gas are not fully established. One hypothesis is that it is monatomic hydrogen, while some other discussions suggest that it might be another form of water. Its origins stem from the work of Yull Brown, an inventor living in Australia. The story circulated that he fueled his car for a year from a garden hose, using 10 gallons of water. In the US, George Wiseman is one of the main inventors in this field. (See George Wiseman's kits and books at *www.eagle-research.com*.) I thus bought an ER50 Kit from George and built the unit, whose picture is shown below.

FIGURE 52. ER50 Brown's Gas Generator from Eagle Research. The unit shows a bubbler unit to generate Brown's Gas for inhalation. The other key accessory is a torch for burning Brown's Gas.

The principle of the machine is electrolysis of water. Unlike normal electrolysis, in which there is just a single cathode and a single anode to convert water into molecular oxygen and hydrogen, the Brown's-gas machine has multiple metal plates between the anode and the cathode. The electrolyte solution connecting the electrodes is a solution of sodium hydroxide (lye). A gas is formed between the plates that is not molecular hydrogen and oxygen. Wiseman believes the gas is another state of water; others talk about it as monatomic hydrogen.

In addition to the prospect of obtaining plentiful energy from water (a subject that Wiseman does not believe is true), there are health benefits described by people, including Wiseman. People may get the health benefits from breathing the gas and drinking the water through which the gas is bubbled. The gas does have 3-6-9 energies, as detected by the 3-6-9 pendulum. The 3-6-9 pendulum has the unique ability to detect GANS (monatomic matter or MagGrav fields).

In my unit, I found that there was a negative-energy quality to the gas when breathing it; it reduced my vibrational energy level. (Reminder: Measuring your vibrational level using the personal wavelength method was described in Chapter 3 of this book, as well as in *The Seeker and The Teacher of Light*). However, by my merely mentally grounding the unit, the deleterious aspect of the gas was eliminated. *Grounding* meant just stating, "I ground the Brown's-gas machine" and imagining a copper wire going into the ground from the machine. I know this sounds weird, but this is the world of subtle energies, which behave by different rules. The grounding completely eliminated the negative effects of breathing the gas using a ventilation tube. The gas induces a very high vibrational state, as indicated when I tested my vibrational level. The Brown's-gas water did not contain 3-6-9 and BG3 energies. However, the gas coming out of the bubbler, which I was breathing, did have 3-6-9 energies.

I did find that if I reused the same water in the bubbler repeatedly (4 or 5 times of 30-minute duration), the resulting Brown's gas no longer increased my vibrational rate (my personal wavelength rotation decreased). Changing the water in the bubbler resulted in the return of the increased vibrational rate. One unsubstantiated hypothesis might be the initial formation of monatomic hydrogen and oxygen that is being breathed, and later—as those concentrations increase in the water—there is increased formation of a new entity of 2 parts monatomic hydrogen with 1 part monatomic oxygen that resembles water but is not water, and that breathing that vapor has a different effect on a person.

You can also hook up the gas produced to a torch to burn the Brown's gas. It has been known that the flame can sublimate tungsten at 6,000 degrees, yet you can pass your hand through the flame. I passed the flame through a variety of objects and metals. Any object that had been passed through the flame would show 3-6-9 subtle energy qualities. I decided to lightly brown a cube of wood with the Brown's-gas flame and compare it to a cube of wood that I had similarly treated with a propane torch. The wood treated with Brown's gas showed 3-6-9 energy, but the wood treated with propane did not show this energy. The gas would vaporize aluminum and cause copper to melt, and the copper that was not melted would glow white-hot and turn the copper black. Currently, I am assuming that this is monatomic copper. Pictures of the wood, copper, brass, and stainless steel are shown below.

  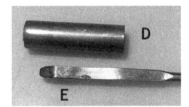

FIGURE 53. Effect of passing Brown's-gas flame on wood, copper wire, brass, and stainless steel. (A) Cube of wood lightly darkened with flame from Brown's gas. Wood now has 3-6-9 subtle energy. (B) Cube of wood lightly darkened with flame from propane torch. Wood does not show 3-6-9 subtle energy. (C) Copper wire heated white-hot with Brown's gas. Tip of wire was melted and entire wire shows blackened copper GANS, which has 3-6-9 subtle energy. (D) Brass tube lightly heated with Brown's gas shows 3-6-9 subtle energy. (E) Stainless steel lightly heated with Brown's gas shows 3-6-9 energies.

I have no doubt that the Brown's gas will produce objects with 3-6-9 energies that can be used to generate fields. Research needs to be carried out to determine other applications of Brown's gas.

Fields Produced by GANS or Ormus Materials

The discussion below describes fields produced by GANS or Ormus materials.

The Fields of GANS and Ormus

Both Ormus and GANS materials possess 3-6-9 energies. The materials themselves do not possess BG3, but they both create *fields* that possess both BG3 and 3-6-9 energies. A major problem for the advancement of the plasma physics of GANS and Ormus has been the inability to detect the fields that these materials create. With the creation of the 3-6-9 pendulum, for the first time these fields can be detected.

In this section, Ormus should be just considered as one type of GANS. Any time the term GANS is used, you can substitute the term *Ormus*.

With the understanding of 3-6-9 energy concepts and the ability to measure this energy easily with a 3-6-9 pendulum, it was discovered that the GANS materials at one location are in constant interaction with GANS at a different location, and that the interaction creates a field that can be measured. Although the GANS materials themselves have only 3-6-9 subtle energies and no BG3, the fields they produce show both BG3 and 3-6-9 subtle energies. The 3-6-9 and BG3 create a way to characterize and visualize these fields.

A simple way to detect GANS fields will accelerate the GANS/plasma science, especially in its applications. The above description of the 3-6-9 and BG3 fields detected with different configurations of GANS or Ormus demonstrate the utility of 3-6-9 and BG3 pendulums to "see" the fields formed by the Ormus or GANS materials. The fact that 3-6-9 energies are present speaks to further evidence of viewing the MagGrav units as being at the heart of creation energy (the torus structure). In the future, there will be many applications of the technology, as we gain knowledge of the principles of how to work with the energies.

Detection of Ormus and the Fields Created by Ormus (GANS)

The following section shows fields created by Ormus when Ormus in vials are geometrically arranged in various patterns. Similarly, fields are shown when flexible tubes containing Ormus are arranged in various patterns.

This is an important new area of study, since researchers in the past have not been able to detect the fields generated by GANS or Ormus, nor the material they created containing Ormus or GANS. With the discovery that any GANS/Ormus material can have 3-6-9 subtle energies, and that they create fields which contain both 3-6-9 and BG3 subtle energies, science has a new way to detect and measure these fields. This should help propel further advances in this important area of study. Any of the other types of GANS should behave the same way as the Ormus in terms of detection by use of their 3-6-9 energies and the BG3 fields they create.

In the figures below, a vial of Ormus is shown as a rectangle. BG3 is shown as circles, and 3-6-9 Is shown as triangles.

Figure 56, below, shows various configurations of Ormus vials and Ormus tubes and the 3-6-9 and BG3 energies. The legend in the description of Figure 56 describes the 3-6-9 and BG3 found. Note that Ormus can be made with inherent BG3 by being given love and blessings during its preparation. This would be analogous to giving BG3 to water by giving it love and blessings. However, the Ormus that I prepared for the purpose of studying its fields and properties did not go through this process, since I wanted to study the chemical/physical properties. Thus, the Ormus I used for creating the fields described had only the 3-6-9 subtle energies, and no BG3.

You will also notice that some geometric arrangements of vials show no BG3 and 3-6-9 subtle energies. In the following section, I compare GANS to disc magnets. With disc magnets, similar fields are produced as with GANS when the magnets assume the same geometric configuration as the GANS vials. If magnets are placed in an unbalanced manner (i.e., an odd number of magnets, or magnets in the center of a geometric configuration) there is imbalance, and the BG3 and 3-6-9 subtle energy can no longer be detected. This explains why GANS vials in certain geometric patterns show no BG3 or 3-6-9.

The above phenomenon also explains the reason for the creation of BG3 subtle energy. When materials are in harmony, there is BG3.

Since GANS or Ormus are MagGrav units of matter (atoms), they are, in essence, miniature "magnets" (or, you could say that magnets have their properties because of their magnetic-gravitational properties). Thus, when you have 4 GANS in a square pattern (i.e., a cube), then the GANS will induce all MagGrav units in the space enclosed by the GANS vials to arrange themselves in an organized and balanced manner. When that happens, BG3 is formed because all MagGrav units are in balance or harmonized (see next section).

3-6-9, BG3, and Fields Produced by Ormus/GANS

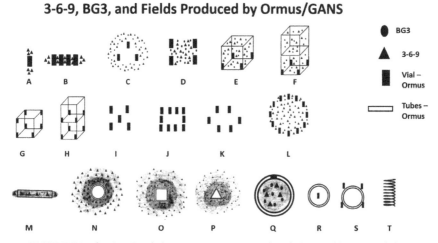

FIGURE 54. The 3-6-9 subtle energies associated with Ormus/GANS and the BG3 in the fields they create.

A. *Single vial of Ormus:* The 3-6-9 field is above and below the vial. No BG3 was detected.

B. *4 vials of Ormus in a row:* The 3-6-9 energy follows the line of vials and extends slightly beyond the vials. No 3-6-9 exists above or below the vials. No BG3 fields were noted.

C. *3 vials of Ormus in a triangular configuration:* No 3-6-9 and no BG3 fields were detected inside the triangle formed by the vials. Both BG3 and 3-6-9 were found surrounding the vials. The fields did not project above or below the vials.

D. *4 vials of Ormus in a square or rectangle configuration:* 3-6-9 and BG3 fields were found within the perimeter of the 4 vials. The energy goes above and below the vials (both 3-6-9 and BG3). No energies were found outside of the perimeter of the 4 vials.

E and F. *Vials of Ormus at the base of a cube or rectangular structure:* BG3 and 3-6-9 fields fill the entire space bounded by the vials of Ormus; the square or rectangle can be just inches, or the size of a house (F: both floors). It was found that the fields continue to build each day. Over a timeframe of weeks, the field strength was 3700 BG3 units in a house. This was determined using a spherical ball on the BG3 ruler to determine BG3 levels of a space.

G and H. *Vials of Ormus at the corners of a cube or rectangular structure* (e.g., 1 set on one floor of a house, and another on a second floor): This configuration resulted in no 3-6-9 and no BG3 detected.

I, J, K. *Vial configurations of Ormus with no BG3 and no 3-6-9 fields detected within and outside of the vials:* No fields were detected above and below the vials. Likely reason: MagGrav fields are unbalanced because of odd number of vials on a side, or presence of additional vial in center.

L. *Vials of Ormus in the circular balanced pattern shown in the Figure:* The 3-6-9 and BG3 fields were within the circle. At the location of the vials, one sees only 3-6-9 and not BG3. The height of the fields was only inches above the circle.

(**M through T.** *Plastic tubes filled with Ormus in water:* Shading to show 3-6-9 field extended beyond BG3 field in N, O, and P array.)

M. *Tube in a line:* 3-6-9 was found along the tube and a little beyond the tube. No BG3 field was detected.

N, O, P. *Tube in a circular, square, or triangular configuration:* The square and the triangle are similar, since they were merely the circle pulled into a square configuration or a triangular configuration. No 3-6-9 and BG3 were found in the center of the configuration. The 3-6-9 field goes out a distance about twice the diameter of the circle and about the same height. The BG3 field goes out a distance about the diameter of the circle and the same height. There is no energy within the perimeters of the square, with 3-6-9 fields going about twice the width of the square or circle, and BG3 going out about the width/diameter of the square/circle. The fields go above and below the coil of Ormus in the circular, square, and triangular configurations.

Q. *Tube in a circular shape, but not connected:* 3-6-9 and BG3 were found within the circle.

R and S. *Mixing Ormus vials and Ormus in a circular tube:* No 3-6-9 and no BG3 fields were detected with the above configurations—the vial in the center or around the circular tube of Ormus.

T. *Tube wrapped around a cylinder multiple times to form a coil:* 3-6-9 energies can be seen only around the coil, nothing extending away from the coil. No BG3 was detected (similar to just a tube filled with Ormus in water).

Using Magnets to Create Fields, and Implications on the Basis of Forming Fields

Magnets can be considered as behaving in the same way as a Mag-Grav unit. Instead of thinking of magnets as having a north and south pole, think of them as having *magnetic* and *gravitational* properties. Single ceramic disc magnets do not have 3-6-9 or BG3. However, when you put 4 disc magnets in a square or rectangular pattern, you find both 3-6-9 and BG3 within the perimeter of the square or rectangle. The level of BG3 and 3-6-9 found were significant—1,200 for both BG3 and 3-6-9. When 3 disc magnets were put in a form of a triangle, BG3 and 3-6-9 were present surrounding the triangle, while there were no BG3 or 3-6-9 within the triangle.

Thus, the geometry of the magnets in forming fields was the same as using GANS or Ormus in forming fields. This is further confirmation that the MagGrav nature of GANS and the MagGrav nature of magnetism have similarities.

A Comparison of Fields Formed by Magnets and by GANS, Using Geometric Arrays

Below are two figures comparing fields: one formed by circular ceramic disc magnets, and the other formed by GANS.

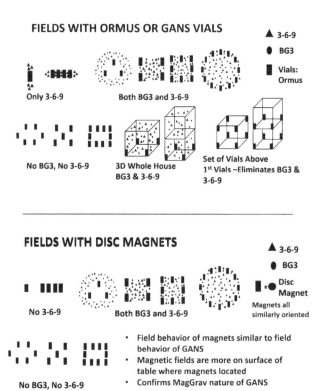

FIGURE 55. A comparison of BG3 and 3-6-9 fields formed by geometric arrays of GANS or Ormus, and those formed by disc magnets. Both form similar fields under the geometric arrays shown. The difference is mainly that arrays formed by magnets were more associated with the surface where the magnets were located, as compared to those formed by GANS, which had more of a 3-dimensional structure.

As you can see, there are obvious similarities in the pattern of the BG3 and 3-6-9 fields formed by magnets and the field formed by GANS. A key difference is that single-disc magnets formed fields that were more two-dimensional (i.e., they did not extend above the table surface where the magnets were placed). GANS have a 3-6-9 subtle-energy quality, whereas single magnets do not. Magnets all have to be oriented in the same direction. In terms of north/south terminology, all magnets on the table surface would be placed either with north facing up, or all facing down.

However, matter is in the form of a double torus (double vortex—male and female, in Walter Russell's terminology). To simulate this phenomenon with disc magnets, I put a piece of paper between 2 disc magnets. The double magnets now behaved the same way as GANS—they now had 3-6-9 subtle energy!

I next created geometric arrays with the double magnets. The array that is in the form of a square or rectangle has both BG3 and 3-6-9, except now the field extends above the table; i.e., the behavior is similar to GANS.

The implication is that GANS has the MagGrav structure, except that the structure is a double torus. Physicist Nassim Haramein came to the same conclusion when discussing how the structure of matter is a double torus.

In order for fields to form, there must be a balance in the geometric configuration of the magnets or Ormus vials. For example, if a rectangle or square is used, the square can be 2 vials (or magnets) x 2 vials, or 3 vials x 3 vials, or 4 vials x 4 vials. If there is not a balance—such as 2 vials x 3 vials, or 3 vials x 4 vials—then no BG3 or 3-6-9 fields are detected. If you place a vial in the middle of a balanced field, such as 2 vials x 2 vials, then no BG3 or 3-6-9 field is detected.

Here are some implications of these observations:

1. Magnets might be better thought of as having their atoms aligned in a similar magnetic/gravitational manner (repulsion-attraction).

2. GANS or Ormus are arranged in MagGrav orientation in a 3-dimensional manner. Each atom is like a disc magnet arrayed in the same direction, but in a 3-dimensional manner.

3. Since GANS forms BG3 and 3-6-9 fields in the space surrounding the vials of GANS, the GANS is inducing a structuring of material in the space to be magnetically arrayed.

4. The GANS itself is magnetically oriented (MagGrav), so it has 3-6-9 properties.

5. BG3 is a harmonizing quality. Geometrically, when everything is in balance, BG3 is present. If all MagGrav material in a GANS field is structured magnetically, the resulting field would be balanced and BG3 would form.

6. The 3-6-9 quality is present, since all elements formed by the GANS in the field are oriented as MagGrav units, which has 3-6-9.

7. The data is further proof of the MagGrav nature of both GANS and the fields they produce.

8. Fields are formed by inducing formation of MagGrav orientation of elements/structures in the space being influenced by the GANS or magnets.

9. BG3 and 3-6-9 fields are formed only when the geometric array is balanced (same number of vials or magnets on a side). This is not surprising, since if the numbers are not the same, this will unbalance the magnetic field.

10. To better simulate matter, if a piece of paper is inserted between 2 disc magnets, the resulting double magnets behave like a double torus, which is believed to be the basis of matter. The double magnets now behave in the same manner as GANS (which is likely a double torus) in forming 3-dimensional fields.

Fields Formed by a 3-6-9 Array of Magnets and GANS/Ormus

I decided to see what type of field is formed by a 3-6-9 array of both magnets and GANS/Ormus. A 3-6-9 array is defined as 3 magnets or GANS vials in the center, 6 magnets or GANS surrounding the 3, and 9 magnets or GANS surrounding the 6. The figure below is the field that is formed by both the magnets and the GANS.

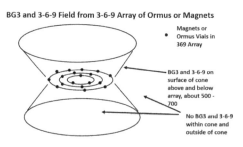

BG3 and 3-6-9 Field from 3-6-9 Array of Ormus or Magnets

Magnets or Ormus Vials in 369 Array

BG3 and 3-6-9 on surface of cone above and below array, about 500 - 700

No BG3 and 3-6-9 within cone and outside of cone

FIGURE 56. BG3 and 3-6-9 fields formed by 3-6-9 array of Ormus or disc magnets. The field forms on the surface of the cone above and below the Ormus. The field is not inside or outside the cone.

Initially, I thought there was no field, since I could not detect a field above and around the array. Only when I went out beyond the array and up could I detect the BG3 field and discover that the field formed on the surface of a cone.

If a magnet or an Ormus vial is placed in the center of the magnet or Ormus array, the BG3 and 3-6-9 field is activated to form the figure below.

BG3 and 3-6-9 Field from 3-6-9 Array of Ormus or Magnets with Addition of Ormus or Magnet in Center of Array

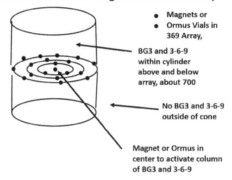

FIGURE 57. By placing a magnet in the center of a magnetic array or an Ormus vial in the center of an Ormus array, the BG3 and 3-6-9 field becomes a cylinder above the array. The field fills the entire cylinder above and below the array.

The two figures above show that the nature of the BG3 and 3-6-9 field changes considerably when a magnet or an Ormus vial is placed in the center of the array. The field changes from being on the surface of a cone to becoming a cylinder that fills the entire cone with BG3 and 3-6-9.

Intuitively, I thought of activating the 3-6-9 array that did not have Ormus or a magnet in the center with just the *intention* of activation, to see if there is a change in the BG3 and 3-6-9 field. Amazingly, the mere thought of activation changed the BG3 and 3-6-9 field to become a cylinder above the array, just as if I had actually placed a magnet or Ormus vial in the center of the array.

This is another illustration of the power of intention and a reflection of who we are.

Another View of Food and GANS

Mehran Keshe believes that humanity may eventually make GANS food, which can be continuously produced, resulting in an indefinite supply of food. The belief is that food in the GANS format is the form best utilized by the body.

I wanted to further understand the nature of food intake—that is, is intake of GANS a key aspect of food digestion? When we eat, is the purpose of our eating to convert food to GANS for optimal uptake by the body? Are there other purposes for cooking, other than making the food tastier? As a result, I decided to carry out a few simple experiments.

The First Experiment (Flax Seed and Lemon Juice)

We know that digestion in the stomach uses bile and acids, and that digestion occurs at body temperature. I thus ground up some flax seed, placed it in a test tube, and then added some salt and a small amount of lemon juice for acidification to a pH of 4.0 (mild acidification). The result: there was no BG3 or 3-6-9. Then the temperature was brought to 97 degrees Fahrenheit. BG3 and 3-6-9 occurred in about 5 minutes.

Experimenting with another flax seed/lemon juice mix, I added a small amount of sodium carbonate, which brought the pH to 4.5. The temperature was brought to 97 degrees F. In this case, BG3 and 3-6-9 occurred within 3 minutes—an acceleration in comparison to the flax seed/lemon juice mix without any sodium carbonate.

Finally, experimenting with a third test tube of flax seed/lemon juice, I added some hydrochloric acid to bring the pH to 2.0. This resulted in an immediate generation of BG3 and 3-6-9, even without bringing the mix to body temperature.

The experiment above demonstrates that the digestion process changes food to produce BG3 and 3-6-9. Increasing the temperature of food to body temperature, use of bile salts, and acidification result in production of GANS, as evidenced by the development of 3-6-9. It is likely that the GANS produce fields that produce BG3. Thus, there may be validity in producing GANS as part of the digestive process of feeding the body.

The Second Experiment (Effect of Cooking on Producing 3-6-9 Energies)

In another series of experiments, I wanted to see the effect of cooking on producing 3-6-9 energies. I thus tested BG3 and 3-6-9 both before cooking and subsequent to cooking. I found that uncooked food (meats and vegetables) did not have BG3 and 3-6-9. However, the cooking process always resulted in producing BG3 and 3-6-9, usually around the range of 1400 level.

The net result of the two sets of experiments is the finding *that digestion helps convert the food we eat into GANS, and that cooking enhances the conversion of food into GANS.* It may be that production of GANS is a vital part of how our body utilizes food.

Summary

In this chapter, you learned about the fascinating world of GANS (Gas in Nano State) and the plasma fields that it creates. GANS is in resonance with 3-6-9 subtle energies. Everything that I have tested seems to be able to be converted to GANS.

From my perspective, GANS may be the basis of all material. Under certain conditions, GANS is released from matter. GANS is detected by 3-6-9, which is indicative that the GANS has the basic torus vortex structure. This is seen in the MagGrav units (description of GANS), which may be the generative gravitational aspect of the torus and the magnetic/radiative portion of the torus in Walter Russell's terminology. GANS is liberated from matter by heat, acids, and alkali. Once formed, GANS can induce GANS formation from other elements.

GANS creates fields, most likely by formatting existing GANS (plasma) in the space where the original GANS is located, by utilizing their magnetic properties. Geometric positioning of GANS results in the creation of fields. The resultant fields have BG3 energies. This may be due to geometric positioning of the GANS/plasma in the field. If the GANS is geometrically evenly spaced in the field because of the magnetic properties of the GANS, then that may induce formation of BG3. Different geometric configurations result in field formation.

These fields may be important for many applications. With the 3-6-9 pendulum, there is now a way to find the location of these fields.

GANS formation may be the essence of digestion. The digestive process appears to be conducive to GANS production from food. Acids, body temperature, bile salts, and cooking all help with the formation of GANS from food.

In the next chapter, we will make use of fields that we form, and show how our consciousness can interact with those fields.

PART VI

Spirituality

In this section, we learn that we have amazing abilities. With a few words, we can create harmony with structures that were previously in disharmony. With words, we can copy or move subtle energies. As Dr. William Tiller showed, intention can modify matter. The channeling of Ra describes human evolution, as we learn that there is only One, as well as the lessons we will all learn as we evolve. This section provides notes on the wisdom of Dr. Karim, as well the channeled messages of SETH. Ultimately, we learn that we are all Creators, and we learn to take responsibility for what we create.

CHAPTER 10

Who We Are—Moving Fields and Copying Subtle Energies/Fields

This is what we have learned so far about moving fields with simply the power of intention and thought:

- When certain BioGeometry tools and 3-6-9 symbols are placed on top of or next to each other, there is interference and the BG3 and 3-6-9 that were present go away. We then learned that simply by inviting in harmony, the interference is harmonized and the BG3 and 3-6-9 energies return. *This is the power of intention and thought.*

- We can take a vial of Ormus in water, which has no BG3 by itself, and by holding the vial in our hands and simply giving it the intention of love and harmony for about 40 seconds, the Ormus vial now has detectable BG3.

- If we put our hands around a cup of tap water for 1 minute and give the water blessings of love and gratitude, the BG3 level of tap water will move from around 200 to 1,200. This is another example of the power of intention and thought.

- You can do this experiment at home. If you don't have a BG3 pendulum, test for 3-6-9 with a 3-6-9 pendulum and you will get the same result—but in terms of 3-6-9 subtle energy rather than BG3.

- We can create detectable fields with vials of GANS or Ormus. Our 3-6-9 pendulums rotate clockwise when fields are present. If we put 4 vials of Ormus into a square pattern, we detect BG3 and 3-6-9 within the perimeter of the vials. The perimeter can be

157

large (e.g., putting the Ormus vials at the perimeters of a house) or small (e.g., a few inches or a few feet apart).

Influencing the Fields

As I was studying the fields, I wondered if it is possible to influence them. To satisfy that curiosity, I physically put 4 vials of Ormus in a square pattern. This generated BG3 and 3-6-9 energies within the square perimeter. I then simply *thought*, "Move the field from within the perimeter to the adjacent area outside the perimeter," and when I measured the field, it was no longer within the perimeter but adjacent to it!!!

I then asked my wife to move the field further away to another location. I wanted to find out whether anyone could do this, even without prior belief that it could work. She thought it was a preposterous request, but she complied. I checked the new location and found that the field had again moved; it was no longer at the original location within the perimeter of the Ormus vials! My wife had no practice or preparation for moving the field; she merely thought and said, "Move" and indicated a new location. The new location had both 3-6-9 and BG3 energies.

To determine whether the Ormus vials were essential to the location of the field, I picked them up—and the field was no longer present at the new location. This meant that the field is present only while the Ormus vials are continuing to generate it.

I experimented further with moving fields. I found that I could take a small field, where the perimeter of the Ormus vials was 10 inches apart, and move the field to a larger location, such as a whole wall. BG3 and 3-6-9 were present on the full wall.

A "Move the Energy" Experiment

Because the Covid pandemic was going on at the time, I asked my friends Jan, Rick, and Joachim to play the "move the energy" experiment on Zoom. They all know BioGeometry and have both BG3 and 3-6-9 pendulums.

Visualizing with the computer screen, they moved the energies from one house to another house, and placed the energies at various locations on the wall without telling the other person the location. The third party would be able to test the different locations on the wall and determine where the BG3/3-6-9 energies were located.

A question then arose: if we could move energies from a field created by a GANS material, could we move energy from a *non*-GANS source? BioGeometry tools such as the harmonizing Cube produce BG3 for a room or house. My friends and I found that, using the BG3 Cube, we could move the energy from a location at one house to a location at another house.

Measuring Other Subtle Energy Qualities

The experiment then expanded to include other subtle energy qualities that we could measure. So I looked at the subtle energy of various items.

In one example, I determined the energy of an essential oil—in this case, lavender (it has horizontal infrared energy)—and moved that energy across the room.

Another experiment was carried out to determine the wavelength of a color using a neutral pendulum, and then moving that subtle energy color quality to another location. The color quality had indeed moved to the new location. This is an easy experiment, which you can try.

The Effect of Moving Fields on the Level of Energy

The next question that interested me was: *How much subtle energy is in a field, and if the field is moved, what is the level of energy in the new location as compared to the old location?* I had already determined that a field containing BG3 and 3-6-9 will activate water by giving it BG3. The questions I wanted to test now were: (1) how quickly will the field that was moved activate water, as compared with the original field, and (2) what is the strength of the field that was moved, as compared to the original field?

The design of the experiment is shown in the figure below, as well as the rate of water activation in the relocated field.

Create Ormus Field of BG3/3-6-9 and Move Field by Intention
Effect of Field on Water

- Place test water to where field was moved
- Place control water to another area
- Test BG3 and 3-6-9 for control and test at various times

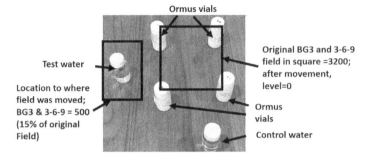

Ormus vials

Test water

Location to where field was moved; BG3 & 3-6-9 = 500 (15% of original Field)

Original BG3 and 3-6-9 field in square =3200; after movement, level=0

Ormus vials

Control water

FIGURE 58. Experimental design for moving an Ormus-generated BG3/3-6-9 field to a new location. The relocated field was then tested for its ability to activate/structure water placed in the new field, as compared to a control vial of water at another location. The original field had a BG3/3-6-9 value of 3200. The new location for the field had a BG3/3-6-9 value of 500, or about 15% of the original value.

BG3 and 3-6-9 Levels of Test Water in Location of Moved Field Compared to Control

Time (Hr)	BG3 and 3-6-9 Levels	
	Control	Test
0	250	250
14	250	700
27	250	800
51	250	1200
120	500	3200

FIGURE 59. The effect on activation of water in a BG3/3-6-9 field that had been moved to a different location by thought/intention. The table shows the kinetics of water activation, and offers proof of the movement of the field from one location to another location. The original BG3 and 3-6-9 had a value of 3,200 within the perimeter of the Ormus vials. The test area containing the moved BG3 and 3-6-9 had a value of only 500. The Ormus field in the original area activates to 3200 within 12 hours

This experiment demonstrates that the relocated field has about only 15% of the strength of the original energy. The relocated field takes about 5 days to activate water to a BG3 or 3-6-9 level of 3,200 (very high), whereas the original Ormus field requires only 12 hours to reach 3,200. The control water was not activated. (The control was located at about the same distance away from the initial Ormus field as the test water in the relocated field.)

This finding proves the relative ease of moving a field from one location to another location. It also demonstrates our innate connection to subtle energy fields and our ability to interact with those fields. Since GANS is monatomic matter, the experiment shows that we interact with this form of matter and/or its fields with intention. We are connected with the matter and its fields. This understanding also offers further information as to who we really are, each one of us—the I AM, our true essence.

CAUTION: GANS or Ormus Vials Around a House May Pick Up Detrimental Energies

Four Ormus or GANS vials placed at the perimeter of a house at the ground floor create BG3 and 3-6-9 energies throughout the house, based on the field it creates. It geometrically harmonizes the house with its distributed fields. The literature on this subject includes a description of placing the vials in the ground around a particular property, indicating that this protected the grounds from a tornado.

Placing Ormus or GANS at Perimeter of House Can Be Dangerous!

Based on the above indication, a friend buried 4 GANS bottles around the perimeter of her house, and this increased the BG3 and 3-6-9 to about 3,200 on the BG3 ruler. I similarly noticed that the energies around my house were increasing to the same extent when I placed 4 GANS vials at the corners of my house. A few weeks after we started this experiment, I checked with my friend to see if the BG3 and 3-6-9 levels in her house were still high, and if they were changing. She detected *no* BG3, whereas there had been plenty of BG3 before. Then she learned that the grounds around the house had been sprayed the day before with a pesticide. Being a BioGeometry practitioner, she tested for the detrimental vertical negative green energy. She found that, due to the pesticide, her house was filled with vertical negative green!!! (That is, the pesticide contained vertical negative green energy.) This could possibly be lethal.

In the early days of radiesthesia, a French scientist, Leon de Chaumery, was using large hemisphere structures to conduct research on energies. He was found dead and mummified. The hemisphere had created vertical negative green, which is very harmful to health and which dehydrates (mummifies) the body.

The *cautionary* point of this tale is that fields from 3-6-9 GANS may pick up environmental energies—and if those energies are detrimental, that energy can spread to the whole house if the GANS or Ormus has been placed at the perimeters of the house. Research is needed to understand these energies. They can be a boon for humankind, but caution is also needed.

Without the 3-6-9 and the BG3 pendulum, a researcher can be blind to the effects of the fields being generated. Since working with GANS and Ormus is a new field, having a 3-6-9 pendulum to detect and measure the subtle energy quality is important. Increasing BG3 in a house or room using GANS is beneficial, but the precaution is that the fields also can pick up other energies that can be harmful.

Another Level of Protection—I AM

Knowing the potential dangers of vertical subtle energies, my friends and I set out to see whether it is possible to protect oneself from ordinary undesirable subtle energies. Using Zoom and the computer, we sent various energies to our friends at different locations. Since we could test for those energies with our pendulums, we had proof that we could successfully send them.

Joachim Wippich (one of the friends involved, and the teacher in *The Seeker and The Teacher of Light*) knew that we can bring our immune system up to 100% by bringing ourselves to the I AM vibrational level. As described in that book, we can test to determine whether we are at the I AM level by simply testing our vibrational level, based on the rate of rotation of our pendulum.

It seemed logical to perform an experiment where we would bring ourselves to the I AM vibrational level, and then send various energies to each other and see if we could detect the energies sent.

The result of the experiment was that the various energies sent to a person did not "stick" with the person. When we tested for the subtle energy that was sent, it was not present on the other person. Thus, by maintaining yourself at the I AM level, you are automatically protected from energies coming from others. Being at the I AM level may create such a high level of energy surrounding you that other energies do not penetrate to your I AM essence.

However, the best protection is to be cautious and not have toxins and poisons within the fields generated by GANS or Ormus.

Structuring and Activating Water by Intention with BG3 Evidence

Since ancient times, people have blessed their food and water, and given gratitude and intentions of making the food beneficial. We also know that water with a high BG3 level is harmoniously structured. Dr. Karim, founder of BioGeometry, has had water that has been harmonized with BG3 tools analyzed in Masaru Emoto's lab. Emoto has demonstrated that when water is harmonized with positive words such as "Love," "Peace," and so on, they form beautiful hexagonal crystals when the water is frozen. (You can see these images in his book, *The Hidden Messages in Water.*)

The tap water in my home had a BG3 level of 200. But I found that by putting my hand around the water and blessing it with love and gratitude, in only 1 minute the BG3 and 3-6-9 of the water went up to 1,200, as measured by the BG3 ruler.

The lesson here is that blessing food and water affects the actual structuring and harmony of the water and food.

The other key lesson is that we have the ability to affect the physical structure of water through our thoughts, as evidenced by the change in BG3 level.

Copying Fields and Subtle Energies

I was listening to a lecture by Dr. Igor Nazorov, explaining how he and Dr. Yury Kronn captured subtle energies by means of gas-discharge tubes with frequency-generator or laser systems with recorders, and imprinting the energies on mineral-enriched water. My natural question was, "Is it possible to use *intention* to move a field, or any subtle energy, onto a computer CD disc?"

I then asked my friends Joachim Wippich, Jan Walsh, and Rick Skalsky to carry out this experiment, along with myself, to move whatever field or subtle energy they wanted. I chose the Aum Array of Array (see figure below) to move the high level of BG3 and 3-6-9 subtle energy onto the CD. Jan and Rick chose to use a BioGeometry field from a BioGeometry tool (e.g., a cube). We were each able to move the subtle energy to the CD. Joachim moved his digital Affirmations on his iPhone onto a CD at my home while we were on a Zoom conference.

If we then moved the tool away from the location to which we had moved the field or subtle energy, the subtle energy returned to the array or cube. Since our BioGeometry tools are useful and we did not want to lose their functionality, we asked ourselves whether it was possible to copy the field or subtle energy with thought/intention, and then embed it into the CD in a permanent manner.

We carried out the experiment. We found that we could still detect the high level of BG3 and 3-6-9 on the original BioGeometry tool or Aum Array of Array. But the high level of BG3 and 3-6-9 could now be detected on the CD. That level of subtle energy remained with the CD for five days, and probably much longer. The digital information on Joachim's iPhone showed an amazing level of BG3 and 3-6-9 subtle energy levels of 20,000 transferred onto a CD with just intention!

The next logical question was whether intention could copy subtle energies onto other types of materials. The experiment I chose to do was to use conscious intention to copy, on a permanent basis, the subtle energies of the 3-6-9 aum array of array onto wood, paper, and stone, as well as a CD. I also wanted to see if the copied subtle energy would activate water, as measured by the BG3 and 3-6-9 levels. The results are shown in the figures below.

Wood, Stone, CD, Paper for Water Activation

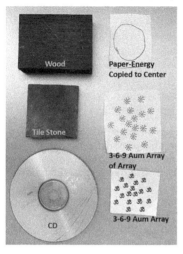 

FIGURE 60. The above figure shows the materials (wood, stone, CD, and paper) imprinted with the subtle energies of the 3-6-9 aum array of array, as shown. The 3-6-9 aum array (3 aum symbols surrounded by 6 aums surrounded by 9 aums) was miniaturized and used to create another 3-6-9 aum array of array. The aum array has 6,400 BG3 units, and the aum array of array has 17,500 BG3 units. A vial of water was placed on each surface and monitored for the kinetics of water activation by following its BG3 and 3-6-9 levels. A vial of water not placed on an imprinted surface was used as a control.

Subtle Energy of 3-6-9 Aum Array of Array Copied to CD, Wood, Paper, Stone Were Still Stable After 42Hours

		BG3 & 3-6-9 Levels			
	Aum Array				
Time (Hr	of Array	CD	Wood	Paper	Stone
0	17,500	17,500	17,500	17,500	17,500
2.5	17,500	17,500	17,500	17,500	17,500
4.5	17,500	17,500	17,500	17,500	17,500
16.5	17,500	17,500	17,500	17,500	17,500
23	17,500	17,500	17,500	17,500	17,500
30	17,500	17,500	17,500	17,500	17,500
42	17,500	17,500	17,500	17,500	17,500

FIGURE 61. To check the stability of the copied energies of 3-6-9 aum array of array, the energy was measured over 42 hours to see if the energy level remained stable. The data shows no deterioration of the energy level.

**Kinetics of Water Activation with 3-6-9 Aum ofAum Array
Copied onto CD, Wood Paper and Stone**

			BG3 & 3-6-9 Levels			
		Aum Array				
Time (Hr	Control	of Array	CD	Wood	Paper	Stone
0	200	200	200	200	200	200
2.5	200	300	300	300	300	300
4.5	200	600	600	600	600	600
16.5	200	1500	1500	1500	1500	1500
23	200	1700	1700	1700	1700	1700
30	200	2600	2600	2600	2600	2600
42	200	3800	3800	3800	3800	3800

FIGURE 62. The table shows the kinetics of water activation as measured by the level of BG3 and 3-6-9 over a 42-hour time period. Results show the same level of activation of water by all of the materials used. One can consider the original aum array of array as a control, since all surfaces caused the same level of activation as the original array.

The data shows that intention can be used to copy subtle energy onto various materials. The copied subtle energy was stable on the materials for the duration of the time tested, showing no deterioration. The subtle energies copied onto the different materials were shown to be able to activate water, as demonstrated by the kinetics of water activation.

Another question that I decided to test was to see if the subtle energies on the various surfaces could be photocopied. I put the wood, CD, stone, and paper onto the surface of a photocopying machine. The photocopied image of the materials also had the same level of subtle energy as the original 3-6-9 aum array of array. You can test the BG3 and 3-6-9 of the photograph and see if you can detect BG3 and 3-6-9.

From my experiments, it is much better to *copy* subtle energy and place it in its new location, rather than to move subtle energy. When you move subtle energy, the energy on the original location becomes affected and is diminished. When you copy the energy, the original energy does not get diminished and can be copied multiple times without getting diminished. And just by *stating* that the copied energy is copied *permanently* onto a new location or medium (stone, paper, or wood, in the above example), the energy remains at the location permanently. It has been noticed that, in certain cases, if the word "permanent" is not used, the energy may diminish over time.

Who We Are and the Work of Dr. William Tiller

In the preceding chapter, I described my work on affecting water with consciousness and intention, and on creating and moving fields and using those fields to activate water. Measurements were made using radiesthesia tools. Throughout this book and in my previous book, *The Seeker and The Teacher of Light*, I have given directions on how anyone can learn to use the tools of radiesthesia to confirm my observations. At this point, many practitioners of BioGeometry and radiesthesia have done so.

For those wanting to use more conventional scientific measurements to prove the effect of consciousness and intention on matter, I will introduce the work of Dr. William Tiller and his colleagues regarding intention experiments affecting the pH of water and of enzymatic reactions. Dr. William Tiller, along with Dr. Walter Dibble, Jr. and Dr. Michael Kohane, authored the book, *Conscious Acts of Creation: The Emergence of a New Physics*. Much of Dr. Tiller's work was carried out while he was a professor at Stanford University. Dr. Nisha Manek did an excellent job of describing his work in her book, *Bridging Science and Spirit: The Genius of William A. Tiller's Physics and the Promise of Information Medicine*.

The first set of Tiller's experiments has to do with using intention to change the pH of water. *pH* is a term used to describe the acidity or alkalinity of water. The "H" in pH refers to the *hydrogen ion* concentration. If the pH is 1, it means this is a strongly acidic concentration. Such an acid concentration will cause serious acid burns.

The pH scale is logarithmic, so a pH of 2 has 10 times lower concentration of hydrogen ions, as compared to a pH of 1. Vinegar (chemically known as *acetic acid*) has a pH of around 4. Neutral water has a pH of 7, which is has a hydrogen-ion concentration a million times less than an acid of pH 1.

pH above 7 enters into the realm of *alkaline* pH. A pH of 14 is very alkaline and will cause alkaline burns. A strong acid would be hydrochloric acid (your stomach makes this acid to aid in digestion, but a prolonged acidic condition can cause ulcers). A strong alkali would be lye (sodium hydroxide); it is the basis of Drano, used to dissolve blockages in drains.

Rather than having a person hold an intention for an excessively long time in order to raise or lower the pH, Tiller decided to use an electronic device that might be able to be imprinted with intention. The device could then continuously be imparting the intention. The concept is not new, as exemplified by the work of Jacques Benveniste, who used a circuit with an electromagnetic input coil and an electromagnetic output coil to capture the energetic fingerprint of ovalbumin in a water solution to transmit it to a vial of pure water. The water picked up the ovalbumin fingerprint, which was then shown to be biologically active (see Tiller's *Conscious Acts of Creation*, page 67).

Tiller's device had an EEPROM (Erasable Programmable Read Only Memory) component, an oscillator component to pick up intention, and a power supply (line voltage or battery). A protocol was developed for imprinting the device, which was then called IHD (Imprinted Host Device). The protocol consisted of 4 persons who were experienced meditators (2 men and 2 women). First they mentally cleansed the environment; then they focused on the prearranged intention (which was first read aloud) for 15 minutes; and lastly they sealed the imprint into the device with a closing intention. Unimprinted Electronic Device (UED) was used as a control. Protection of IHD and UED was carried out with a fine-mesh copper Faraday cage and aluminum foil. The tool used to measure pH was a sensitive pH meter with electrodes.

The many experiments carried out by Tiller and his group were very definitive. It was found that the pH could be moved up or down by as much as 1 pH unit upon exposure to the IHD, when compared to the control UED. This proved the concept of imprinting an electronic device with intention, and measuring its impact on water by changing the pH up or down. Below is a figure from Dr. Nisha Manek's book, *Bridging Science and Spirit*.

In addition to studying the effect of intention on pH, Tiller's group decided to investigate the effect of intention on the reaction rate of an enzyme. The enzyme they chose was alkaline phosphatase. This is an enzyme that often is tested in the clinical lab as an indicator of liver function. An increase of alkaline phosphatase is indicative of a liver problem.

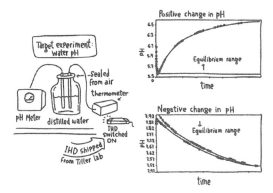

FIGURE 63. Increase and decrease in pH, based on intention imprinted in the IHD.

As with the pH experiments, an IHD was programmed with intention to increase the reaction rate, and a UED was used as the control. The results clearly showed that the IHD with intention to increase reaction rate of alkaline phosphatase indeed showed significant increase in enzyme reaction rate. This is shown in the figure below.

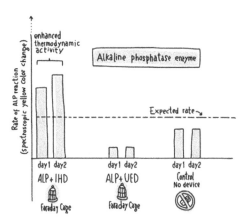

FIGURE 64. Effect on reaction rate of alkaline phosphatase (ALP) upon exposure to IHD, with intention to increase reaction rate, as compared to UED with no intention.

A third experiment carried out by Tiller's group was to explore the effect of intention on the rate of maturation for a living organism: the fruit fly. The fruit fly has a well-defined life cycle of 14 days from egg to larva to adult. Tiller's intention read: "To increase the ATP energy ratio to inactive energy such that the larval development time to adult is 25% shorter or better."

(For those not familiar with the term "ATP," it stands for *adenosine triphosphate*. The molecule is the body's way to store energy and release it when needed by its conversion to ADP, or *adenosine diphosphate*. Thus, the ratio Tiller is referring to is: ATP to ADP. Having more ATP results in greater ability to have energy to carry out the processes of maturation and life.)

The net result of the fruit-fly experiment, using IHD and UED, was that the maturation time did, in fact, decrease.

In Tiller and Manek's description of the effect of intention on pH or enzyme-reaction rates and fruit-fly maturation time, they report on what happens in D-space as compared to R-space. *D-space* (Direct Space) is described as the normal space in which we live. *R-space* (Reciprocal Space) is the mirror image of D-space; mathematically, it is 1/D-space. R-space is conditioned "higher energy" space and is affected by emotion. It is the excess energy which can be accessed in R-space that causes the effects seen in D-space. These theoretical models can be explored further in both Tiller's and Manek's books.

Tiller's work also discusses the Second Law of Thermodynamics, which is concerned with the concept of Free Energy and Entropy. It is the free energy that allows work (i.e., free energy) to be done. Entropy is a component of the equation for free energy. By decreasing entropy, work increases. Life and intention—which can translate to information—would increase free energy by decreasing entropy.

We can view this in terms of equations. The key equation for the Second Law of Thermodynamics is:

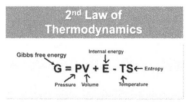

$$\Delta G > 0 \implies \text{Work}$$

FIGURE 65. Second Law of Thermodynamics

If entropy (breakdown of matter) is a larger number, then free energy becomes small. Life processes reduce entropy (e.g., eating food to generate ATP to carry out reactions to enable life). Since intention changes matter, as shown by the experiments performed by Tiller, then intention is a creation process in that it provides information that lowers entropy. The second equation in the figure above indicates that if the change in Gibbs free energy is greater than zero, there is energy to accomplish work.

From my perspective, Tiller's experiments offer examples of intention affecting physical matter. This is further evidence of who we are. In essence, we are acting as creators when we are affecting matter. The mechanism may be via affecting R-space, as described by Tiller. From my data, intention may be affecting subtle energies in fields, which may be similar to Tiller's description of R-space.

Summary

The purpose of this chapter has been to give further evidence that thoughts and intentions have very material effects.

We can change the structure of water with our intention, and we can measure that change. We can harmonize two structures—a BioGeometry tool, such as a cube, and a 3-6-9 symbol (e.g., aum symbol)—which causes nullification of their subtle energy qualities (BG3 and 3-6-9) by merely the thought of bringing in harmony. This brings back BG3 and 3-6-9, which can be measured.

We can ask subtle energy fields to move (BG3 or 3-6-9), and the fields move to new locations. We can then test that the energies have moved by the effects of those energies on water. The energy at the new location will affect the level of BG3 of a test water in the new field that we created.

Similarly, we can ask subtle energy such as BG3 or 3-6-9 to be copied, and then copy that energy onto another material on a permanent basis. Examples of energies that were able to be copied included stone (tile), wood, and paper. The copied energies were able to activate water at the same rate as the original item that was copied.

Our experiments also indicate that when we are at the I AM vibrational frequency, there is protection from outside sources of subtle energies.

The experiments of Dr. Tiller, summarized by Dr. Manek, demonstrate the effects of intention in the material world by affecting the pH of water, rates of enzymatic reactions, and rate of maturation of fruit flies.

The implication of these experiments is that we are all creators, and that we create with thought.

CHAPTER 11

Ra and The Law of One

I find it worthwhile to offer you another view of the story of how we are all connected—one that has to do with the afterlife and dimensions beyond the earth.

Channeling is one way that we are given evidence of our connection to those in the afterlife, and of the truth that we are eternal beings. Channeling is also a way for communicating information from entities with a message, whether they be spirits or extraterrestrials. Some individuals, called *channelers*, have the ability to receive messages, either from those who have died or from entities desiring to communicate their messages. This may occur by the entities or spirits speaking to the channelers and then the channeler relaying the message to living people, or by the entities or spirits actually coming into the channeler and speaking through the channeler's voice. (I wrote about this in *The Seeker and The Teacher of Light* and will go into it briefly in Chapter 12 of this book.)

My two good friends Jeanne Love and Regina Ochoa are wonderful channelers who worked closely with the deceased astronauts of the space shuttle Challenger, as well as of the Columbia space shuttle. (Both stories are described on the website, *https://challengercc.org*.) The Columbia astronaut channeling was carried out in my house. Afterwards, as I was transcribing the recordings of the channeling session, astronaut Sally Ride decided to give me a direct experience of channeling and came through to me in her own voice. I tell this story as my direct evidence of the validity of channeling to obtain information that gives evidence of the perpetuation of life beyond the physical.

I want to share another very remarkable channeling, which is described in *The RA Contact: Teaching The Law of One*, Volumes I and II. The teachings in *The Law of One* are reminiscent of Joachim Wippich's teachings of I AM. In *The Seeker and The Teacher of Light*, we learned that God or Source is the essence of everyone. Our essence is I AM, and when we say "I AM Everything," that describes each of us because there is only One and we are part of the One. We think we are separate because we have each been given Free Will, which gives us diversity. It is our ego and senses that tell us we are separate. But when we realize that we are I AM and that everything and everyone is I AM, we understand our essence, and the Law of One makes sense.

The general story of the Law of One is worth telling. This chapter is not meant to cover the breadth of teachings, since the two volumes of the book are over 1,000 pages. But I can describe pertinent information that gives you the basis of the books, as well as a rudimentary definition of the Law of One.

The channeled message is from Ra. Ra can be described as a *social memory complex*, where memories, thoughts, and experiences of every individual are known to the entire complex. Carla Rueckert is the channeler of the information. Donald Elkins is the questioner of Ra in the channeling sessions. Jim McCarty is the scribe for the recorded channelings. The channeling occurred between January 15, 1981 through March 15, 1984, in 106 sessions. The information gives our history, who we are, and valuable teachings on many subjects.

The ancient Egyptians were visited by Ra, also known as the sun god. Ra came to Earth embodied in a human form for the purpose of teaching humanity about the Law of One. The Egyptian civilization had evolved sufficiently, and Ra's thinking was that the Egyptians were ready to understand the Law of One. However, the message got distorted, and the royal elite reserved the message only for themselves. As a result, Ra removed themselves. However, they felt compelled to correct the distortions the Egyptians had made of the Law of One. This was one reason why they returned later to give the channeling information to Rueckert, Elkins, and McCarty.

The second reason for returning to teach the Law of One is that Earth is now nearing the end of a major cycle of evolution. At the end of the cycle, many of those who are ready will move up the evolutionary stage to what Ra calls "the fourth density." This is the "ascension" that is commonly discussed in the general literature (the end of the Mayan calendar cycle). Humanity is currently in the *third* density.

Ra, however, is in the *sixth* density. They have evolved through their various densities. Their third density was on Venus, 2.6 billion years ago. They have long since moved from Venus. They consider themselves a "social memory complex," as described above. As Earth moves through the fourth density, we will similarly begin the transition to a social memory complex.

The Law of One Cosmology

The following excerpts are taken from the Introduction to *The RA Contact: Teaching The Law of One*, Volume I.

As stated by Ra:

> *In truth there is no right or wrong. There is no polarity, for all will be, as you would say, reconciled at some point in your dance through the mind/body/spirit complex which you amuse yourself by distorting in various ways at this time. This distortion is not in any case necessary. It is chosen by each of you as an alternative to understanding the complete unity of thought which binds all things. You are not speaking of similar or somewhat like entities or things. You are every thing, every being, every emotion, every event, every situation. You are unity. You are infinity. You are love/light, light/love. You are. This is the Law of One.*

In this book, Ra brings us face to face with the same basic truth that has been reported by mystics from all quarters of the world throughout the ages: the astounding realization that the One Infinite Creator is within us and is within everything, everywhere. In fact, the Law of One asserts that there is nothing that is not the Creator; there is nothing that is outside of this underlying unity. Ra reports that the Creator has made the infinite creation out of Itself for the purpose of knowing and experiencing Itself.

This "intelligent infinity," as Ra calls it, generates out of its own being the galaxies, stars, planets, entities such as ourselves, darkness and light, love and fear, every shade of meaning and experience, every mode of thought and activity, and everything else real and imagined on every plane of existence. And It has endowed each and every seeming portion of this creation with a foundation of free will: the capacity to learn, to grow, to intend, to adapt, to make evolutionary choices, to chart a return path of experience to the Creator.

As we travel on our spiritual journeys, we exercise free will, choosing to gradually know ourselves more clearly, and sooner or later we grow into unity with the One Creator. As all of the infinity of entities in the infinite creation travel this path, the One Creator comes to know Itself in ways that are unimaginable and endless through every free will choice that is ever made by each portion of the creation.

Travel Through the Densities

The book further reveals that:

The journey that each soul takes, according to Ra, moves through an infinite system of "octaves," each octave divided into seven ascending densities (or concentrations) of light. In the **first density** *of our current octave, fire and wind teach earth and water to be formed in such a way as to produce the foundation for subsequent biological life.*

The **second density** *is the level of consciousness inhabited by bacteria and single-celled organisms in the lower stages to plants and animals in the higher stages. This density's lessons involve transforming from the random change of first density to a more coherent awareness that facilitates growth and directed movement. As entities progress through the second density, they begin to strive toward the next density of self-consciousness;*

and as the spirit complex becomes awakened, graduation to the third density becomes possible.

Earth and its human population are currently approaching the end of the third-density cycle, according to the Confederation. In this **third density,** the density of choice, we have a more highly developed self-awareness that includes the mind, the body, and, for the first time, a fully activated spirit. The function of this density is to polarize our consciousness and to choose our form of love, our form of service.

On one end of the spectrum of polarization is **service to self:** an exclusive love of self which rejects universal love and seeks to control, manipulate, exploit, and even enslave others for the benefit of the self. On the other end of the spectrum is **service to others:** a love of not only the self, but of all other-selves. Service to others seeks and embraces universal, unconditional love, sees the Creator in all things, and supports the free will of all. Our lived lives are not so black and white, however, as we strive toward either end of the spectrum of polarity in consciousness.

In congruency with various wisdom traditions of Earth, Ra communicates that we are moving toward a "new age," or what Ra would call a harvest, to the **fourth density** of love and understanding. This is where the social memory complex is born, where thoughts become things, love becomes visible, and the positive and negative polarities separate from each other to inhabit environments more suited to their respective and divergent courses of evolution.

The **fifth density** is the density of light, wherein wisdom becomes the focus and criterion for graduation to the next density.

The **sixth density** balances and unifies the love learned in the fourth density with the light (wisdom) learned in the fifth density, and produces a power to serve others that is more effective than that of love or wisdom alone.

*The **seventh density** reaches a realm of experience even more difficult to describe. According to Ra, it is the density of "foreverness," and here we begin to move into total harmony with the One Creator.*

*The **eighth density** represents the complete coalescence of all of the creation with the One Creator, and can be viewed as the first density of a new octave, similar in arrangement to the notes on a musical scale. The fruits of this octave will eventually give birth to another octave of densities, whose fruits will give birth to another octave of densities, and so on, infinitely.*

Commonality between *The Seeker and The Teacher of Light* and The Law of One

In *The Seeker and The Teacher of Light*, we recognized that our essence is I AM. The essence of *everyone* is I AM. I AM is Everything and Everybody. The Creator is within us and experiencing through us. In The Law of One, there is only the One Infinite Creator. Thus, the core message of The Law of One resonates with the core message of *The Seeker and The Teacher of Light*.

Even the term "Seeker" resonates with what we are teaching. Ra said:

The seeker seeks the One. This One is to be sought, as we have said, by the balanced and self-accepting self, aware both of its apparent distortions and its total perfection. Resting in this balanced awareness, the entity then opens the self to the universe which it is. The light energy of all things may then be attracted by this intense seeking, and wherever the inner seeking meets the attracted cosmic prana, realization of the One takes place.

There Is Much Valuable Knowledge in the Ra Materials

Having read summaries of the Ra material in the past, I recently listened to the full version of *The Ra Contact* on Audible. This prompted me to buy and read the full Volumes I and II to get the details of the channeling.

The Ra materials give us a history of humankind and an under-standing of other beings who are also of the mindset of Service to Oth-ers. We also encounter beings who are oriented to Service to Self, such as those from Orion, who are well aware of humanity and try to convert us to Service to Self, for their benefit. We learn of the Wanderers, who incarnate on Earth to help us grow in a more loving Service to Others. We learn about the wars that destroyed those from Mars and much of their planet, and how they were helped to incarnate on Earth. We learn about the building of the pyramids and their purpose. The book con-firms the teachings in this book and the books by Walter Russell when he describes creation of matter, including our physical body, as being the result of the rotations of light.

As an illustration of the teachings of Ra on the nature of creation, I will summarize some statements from the Questioner (Don Elkins) and Ra in Session 29.12. Here, Ra has been explaining the elements of cre-ation and the formation of space/time and manifestation of awareness or consciousness. The Questioner asks Ra for clarification:

Questioner: "As the Love creates the vibration—I will make this statement first. Let me say, I believe that Love creates the vibration in space/time in order to form the photon. Is this correct?"

RA: "I am Ra. This is essentially correct."

Questioner: "Then the continued application of Love—I will assume this is directed by a sub-Logos or a sub-sub-Logos—this continued application of Love creates rotations of these vibrations which are in discrete units of angular velocity. This then creates chemical elements to our physical illusion and, I will assume, the elements in the other, or what we would call nonphysical, or other densities in the illusion. Is this correct?"

RA: "I am Ra. The Logos creates all densities. Your question was unclear. However, we shall state the Logos does create both the space/time densities and the accompanying time/space densities."

Questioner: "What I am assuming is that the rotations, the quantized incremental rotations of the vibrations, show up as the material of these densities. Is this correct?"

RA: "I am Ra. This is essentially correct."

The above statements are very analogous to Walter Russell's description of matter creation, as previously explained in this book. It is also interesting to note that Love is an integral part of the creation process.

The Ra Contact: Teaching the Law of One, Volumes I and II, is a must-read book for those wanting a deeper understanding of our nature and the nature of the universe. There is an Audible version of the two volumes. You also can download any and all of the book and channeling sessions at *https://llresearch.org/home.aspx*. The current chapter is only an introduction to the depth of information found in *The Ra Contact*.

Summary

The Law of One is an amazing set of channelings from Ra. The channeling gives us the basis of who we are—that there is only One. That story is similar to the concept of I AM described in this book and in *The Seeker and The Teacher of Light*. The Ra channelings tell of our evolution to the fourth density in the not-too-distant future, and describe our abilities and life in that density. The important key to graduation to the fourth density is being in the state of Love and a mindset of service to others. It is interesting to note that matter creation, as described by Ra, is similar to the process described by Walter Russell noted in this book.

CHAPTER 12

On Spirituality and Science

Science Means Making Data-Based Observations

Western science is based on conducting experiments and making *observations* (data) to test the correctness of a hypothesis; or (going in the opposite direction) a *hypothesis* may be developed to explain existing observations. If there is enough data, then the hypothesis can graduate to a *theory* to explain the observations.

The Seeker and The Teacher of Light shared many observations indicating that we are not just the body, and that "life" goes on after death. Among the variety of observations discussed were:

- Near death experience (NDE).
- Out-of-body experience (OBE).
- Communications through channeling.
- Electronic voice phenomenon (EVP).
- Instrumental trans-communication (ITP).
- Quantum physics (matter can be a particle or a wave, and is observer-dependent).
- That much of the brain can be missing (hydrocephalus) and yet the person can be fully functional.
- Intention can affect random-event generators (PEAR experiments).
- Energetic auras surround us.
- We have innate abilities such as healing, remote viewing, and dowsing.

There is an abundance of data in these areas. However, sometimes Western science may ignore the data, since it does not fit this paradigm's core beliefs or the standard model of data (such as measurements by known scientific instruments).

The Scientific Criterion of Reproducible Data

This book has presented data indicating the existence of innate abilities that we all possess. Although the data is reproducible (a criterion typical of Western science), it may be dismissed by mainstream science because of unfamiliarity with the tools being used–in particular, the pendulum.

The principle of the pendulum is that it measures wavelength and amplitude through the scientific principle of *resonance*. Detection of resonance takes place through the observer. Anyone trained in the field of radiesthesia or BioGeometry will be able to detect and make measurements through the use of the pendulum tool. Indeed, most people will be able to use the tool if given some training. Principles of how to detect frequencies and amplitudes using pendulums are offered in this book.

You Can Do the Experiments Offered in This Book—Scientifically

Because of all this, I invite you to conduct the experiments described in this book, and through your own experiments to prove to yourself the truth of what is written here. The fact that many people can detect energies and get the same results should satisfy the scientific criteria that *the data is reproducible.*

This book has covered a variety of observations and data. I'll repeat some of them here for your benefit:

- A harmonizing energy quality called BG3 can be detected and measured quantitatively.
- An energy quality associated with the torus structure called 3-6-9 can be detected and measured.
- Water activated with BG3 and/or 3-6-9 can be measured quantitatively.
- Creation symbols associated with the torus structure have 3-6-9 and BG3 energy qualities.

- BioGeometry tools (e.g., cubes, L-90 pendant) have BG3 qualities that can be quantitively measured.

- The direction of pendulum rotation when either BG3 or 3-6-9 is present is a clockwise rotation.

- Many BG3 tools and a 3-6-9 symbol (e.g., torus structure, Sanskrit aum, bagua, yin yang symbol) will interfere with each other, nullifying their ability to be detected. The simple intention to harmonize the energies is sufficient to bring back the energy qualities. A yin yang symbol also will harmonize the energies.

- Ormus and GANS materials, as well as magnets, create 3-6-9 and BG3 fields. These fields can be detected and can be moved or copied to other locations with intention. The 3-6-9 and BG3 subtle-energy qualities can be detected at their new location, even at a distance of many miles.

- Water placed at a location where a field has been moved can be activated to higher levels of BG3 and 3-6-9, which can be measured quantitatively.

- Materials in the process of decay (entropy) will cause a pendulum to spin counterclockwise. Materials in the process of growth or gaining more structure (creation) will cause a pendulum to spin clockwise.

- Foods, supplements, and drugs that are in resonance with us (i.e., are positive for us) cause a pendulum to spin clockwise. Foods, supplements, and drugs that are not in resonance with us (i.e., are detrimental to us) will cause the pendulum to spin counterclockwise.

- As described by Walter Russell, matter *creation* has clockwise centripetal spin, and matter *decay* causes centrifugal counterclockwise spin.

- Our vibrational energy can be measured by the speed or rotation of pendulums at our personal wavelength.

- Our vibrational energy level increases when we remember that our essence is I AM.

- We can bless water and thereby cause the BG3 of the water to increase, which can be measured quantitatively.
- What we create with words, thoughts, or symbols has power and can be reproduced. (E.g., a symbol that we activate to have BG3 or 3-6-9 can be photocopied, and the photocopy will then have that subtle energy quality.)

I consider the above data and observations to be scientifically verifiable, since most people, with just a little training, can perform the experiments and obtain corroborating data.

Data and Spirituality

In a moment, we'll consider how the data fits in with the concept of spirituality. But prior to that, l want to describe how I view spirituality.

To me, growth in spirituality is an educational process involving learning who you are and what that means. It is also the realization that you are spirit living in a body. Eventually, you learn that your essence is I AM, and that even this has various levels of understanding associated with it.

Here are some of the lessons I have learned in my spiritual educational process:

- There is only One. Everybody and everything is part of the One.
- All created matter is created in the same way. It is light in a centripetal clockwise vortex.
- All created matter decays in a centrifugal counterclockwise vortex.
- Creation and decay is a never-ending cycle; what decays comes back in a repeat of the creation-decay cycle.
- Our consciousness, or our essence, is eternal.
- We are all a thought form of the Infinite Creator/Intelligent Energy, experiencing everything. In our evolution, we eventually merge back into the One. We are each the I AM having experiences.

- We can view ourselves as Intelligent Energy. There is no separation—we are One.
- In our current incarnation, we are learning the lessons of Love, Balance, and Harmony—our Oneness of I AM.
- We learn that we are Thoughts, and we create with Thoughts and Words. We also learn that our Thoughts and Words have power, so we want to avoid statements that give reality to such things as fear, anger, diseases, and so on, because we have the ability to create with those thoughts.
- We are each given Free Will.
- With Free Will, we can choose to be of Service to Others or to be Service to Self. Most of us will choose Service to Others and be a loving person. We learn from the polarities that show up in Service to Others vs. Service to Self. In Service to Self, we may not care if we enslave others or cause wars that would benefit mainly the Self.
- I AM can also be thought of as Intelligent Energy, Acknowledgment that we are eternal, and Memory of all of our experiences.
- In our day-to-day experiences, we learn not to judge, we learn to love, to forgive, and to be thankful.

Since we are all One with God/Infinite Creator, I have come to believe that we are all Creators. Being Creators, we need to ask the question, "*How* should we create?" Since everyone has Free Will, polarity in thoughts will occur. There is a need to acknowledge the range of thoughts on all sides. Then, if everyone is willing to listen, solutions will be found that will be a balance. In that balance, there is harmony. In Creating, there is the need for Love, Harmony, and Balance.

Some Spiritual Lessons

I thought it would be good to mention some thoughts and lessons from Dr. Ibrahim Karim (founder of BioGeometry), SETH (Spiritual Energy for the Transformation of Humanity—oversoul), and channelings of Jeshua Ben Joseph (Christ).

Ibrahim Karim (BioGeometry)
From my notes during a lecture by Dr. Karim in May 2021:

The Creator created.

In our creation, we are given Free Will. We are also given the five senses to gather information, and a memory bank in which to store it. This allows the creation of a personality—the ego. Ego only sees the five senses, and its identity is built to serve only the five senses. But the five senses are limited and cannot see the whole. Ego does not see 99% of the information and has only 1% of the information (e.g., the body is run by nature and not by the ego). The world is seen through the eyes of the ego.

As humans learn and see through the heart and inner reflection, the understanding of the beauty of universal consciousness arises. We come to understand that the dualities we perceive are merely a matter of balance and imbalance. The distortions and dualities (our perceptions of good and bad) given to us by Free Will are necessary in order for us to learn the lessons of balance. It is often not about "good or bad," but about creating balance with the perceived imbalance.

We are given tools to help us evolve. As we learn to understand nature and the nature of creation, we learn that there are archetypes from higher dimensions that are used in creation. The spiraling energies in our chakras connect to these higher dimensions. We learn harmony through the translation of information from the right and left brain and the heart.

The key concept that Ibrahim teaches is the law of daily action. This is what creates "excellence of action" in all activities that we perform. Excellence of action connects us to the universal mind.

Concerning the afterlife, Ibrahim's view is that its form is created from the quality of our actions while we are alive. Our judgments of others can create problems, whereas our acceptance of others produces serenity. Separation creates tension. He says: "If we accept all others, we bring those qualities to the other side to make it heaven. Life in the physical dimension creates the other world. Our excellence of action here creates the greater perfection of the afterlife."

SETH (*Spiritual Energy for the Transformation of Humanity—Oversoul*)

From my notes during a channeling session by Jeanne Love, May/June 2021:

> *SETH is an oversoul who has enjoyed many lifetimes on Earth. His name stands for Spiritual Energy for the Transformation of Humanity. Channelers Jeanne Love and Regina Ochoa have had many interactions with SETH. He helped oversee the channelings of the Challenger astronauts with Jeanne and Regina over 30 years ago, when the Challenger space shuttle exploded shortly after launch (see https://challengercc.org or the link from https://fmbr.org). You may know of SETH through various books by Jane Roberts on the many channelings she carried out with SETH (e.g., SETH Speaks).*

> *Through Jeanne Love, SETH gave a "post-graduate" course for a select group of persons. I was fortunate enough to be in that group. Although I won't summarize that course, I thought it would be good to offer some pearls of wisdom from SETH for those ready to hear about the material. (For SETH's many lessons, read the Jane Roberts books.) The following notes do not cover the many volumes of information channeled by Jane Roberts, but they offer much wisdom.*

I appreciate SETH's wisdom, and I appreciate Jeanne bringing SETH's thoughts and words to me and to you—such as:

- Many realities exist simultaneously. These other realities are like a rubber-band ball. Your reality is one of those rubber bands. Your consciousness can move to other bands for you to learn from those other realities.
- The energy that is the soul is created for knowing itself. God is the energy that flows through us and everything.
- Every experience is spiritual. Everything is of your creation.

- Your Words are everything. When you say you are "tired," that is what you become. When you state a fear, you are embracing that fear. Therefore, monitor your words. Be impeccable with your words. Send unconditional love and support.

- Everything is spiraling, so don't expect static perfection. Everything just is.

- You can manifest things. You can be in the field of love.

- How you perceive is what you send into the world. Choose the qualities of your thoughts that you live with every day of your life.

- We are connected with all things created. Our souls are connected, just like all parts of a tree are connected. However, a vast part of humanity is lost and they do not remember their connection to each other and to the planet. We are subjected to propaganda that we are separate—resulting in many lost souls. We can help others to reconnect by means of our right words and thoughts. Bring in the thoughts of connection, love, harmony, and our oneness.

- It is not that we need to shift all paradigms to change the world. Work with the smaller paradigms that can work toward the light. As you raise your frequency, it goes into the webbing of the whole system. It is like increasing the wattage of your bulb to make it stronger.

- You are here to experience, and to remember yourself and the unity and the Allness of yourself.

- The ego is the part of ourselves that allows us to take care of ourselves. We need to integrate it with the other parts of ourselves.

- We need to clean up our language concerning "good" and "bad." We are all programmed with our preconceptions. We see that in relation to foods—e.g., "Don't eat fried food." Instead, we can eat in love and nourishment, and still lose weight. You determine what is right for you. Find your joy. Nourish yourself and others.

- Nurture and be nurtured; give and receive.

- Be connected to yourself and constantly re-evaluate the best way to serve your evolution. Don't get "stuck"—you have divine guidance, if you listen to it.
- You are eternal and can access other vibratory fields, especially through meditation.

The Channelings of Christ

There are a number of phenomenal channeled books where Christ (Jeshua Ben Joseph) comes through a number of channelers.

The most widely read of the channelings is *A Course in Miracles (ACIM)*, channeled by Dr. Helen Schucman and first published in 1976. Helen describes the channeling as "inner dictation" from Christ. The main message of the book is that there is no separation between us and God. The organization's main message is "to awaken us to the truth of our oneness with God and Love." (*www.ACIM.org*)

Another major work is *The Way of Mastery*, the channeling of Jeshua Ben Joseph by Jayem, published by the Shanti Christo Foundation. The main points in the channeling, according to the Way of Mastery website (*https://wayofmastery.com/jeshua/*), include:

- All beings are a spark of Light (soul) at One with God and remain eternally as they are created to be.
- We are each fully responsible for how we use the Light to choose and thus attract precisely what we experience, and this power of choice is the most fundamental aspect of the radiant and sovereign soul.
- God is Love, and Love does not condemn; there is no judgment.
- All beings are to be forgiven and supported (if they are willing) to transform their orientation from fear to Love.
- Only when we forgive and then find ways to serve this "Atonement" may we truly awaken to a full enlightenment and enter Mastery.
- Forgiveness is not an intellectual gesture, but a radical transformation of one's being, best encapsulated in the Aramaic term, *washwoklan*: To return to Original Wholeness.

- While our Union and Reality is unchanged, unchanging, and unchangeable, we have used our power of choice to imagine and create the experiences that flow from a belief that we have achieved Primal Separation.

- This is the true meaning of "hell." Though often hidden, Separation engenders a Primal Guilt, the "dysfunctional" sense of our inner being that gives birth to the myriad forms of psychological and behavioral sufferings of humanity.

- Heaven, like hell, is a potentiality available to all beings willing to enter and go through the purification, healing, and waking process of enlightenment. Heaven is as equally "here" as in any of the infinite realms of Light beyond this world.

- Sin is not a moral failing but a resistance to knowing and living this Light.

A key theme in the teachings is our oneness with God, and that our problems stem from our belief in being separate from God. Enlightenment is the realization of our oneness with God.

Recently I have been attending the channelings of Yeshua and Mary Magdalene by Anthony Compagnone with Renee Blodgett (*https://bluesoulearth.com/*). What I found most interesting is that Yeshua's key message is that people understand who they truly are. In his terminology, the key is in stating "I AM who I AM." In other words, his message is the same as the message taught by Joachim Wippich as to your true essence, I AM.

Common Ground

There is a common thread throughout all the material presented in this book. It is the same thread described in *The Seeker and The Teacher of Light,* as well as in the various chapters of this book. That thread is: There is only One.

On a personal basis, the common thread is *I AM*. As our I AM essence, we are all a reflection of the Infinite Creator, created in order for the Creator to experience Itself. All matter is made of the spiraling light of the Creator. All matter is imbued with a consciousness, or intelligent energy, of the Infinite Creator. As creation evolves, greater awareness develops.

We are given Free Will, which allows us to develop in many ways, enabling us to have different experiences. As humans, we have evolved beyond the level of just awareness; we also have the ability of abstract thinking.

In our growth, we learn the lessons of Love. To experience and to allow the ego and personality to develop, we are given our five senses and memory. From our Free Will, we choose what we want to experience. From ego and judgment, we develop the concepts of duality, of good and bad.

We may develop personalities, which may be inclined to Service to Self or Service to Others. We forget that we are One, that we are all the Infinite Creator experiencing All. Our ego and personalities lead us to believe that we are separate, and we behave accordingly. Our definitions of "good" and "bad" are based on our perceptions, which can change. Our perceptions may result in all forms of disagreements/conflicts. We are not separate, and we can realize that our differences are best handled by utilizing the principles of Love, Harmony, and Balance.

As we develop and evolve in our maturity, we learn that what we do and think need the elements of Love, Harmony, and Balance. We are all Creators. We learn that everything is thought and light. When a majority of humans have learned the lessons of Love, Harmony, and Balance, we will have advanced to the next level in our evolution.

The common threads described above are seen in the teachings of I AM by Joachim Wippich. They also are reflected in the teachings of Ra, Dr. Ibrahim Karim, and SETH.

Summary

The science described in this book says that everything is created in the same way, with spirals of light, and that there is consciousness within all that is created.

With our ability of abstract thinking, we can realize that we are Creators, and that we create with Words and Thoughts. Because of our misconception of separateness and lack of understanding of our Oneness, we often do not realize the power that we exercise when we create with Words and Thoughts. We need to understand that we are Creators and to behave accordingly. We need to create in Love, Harmony, and Balance if we want to avoid the conflicts that we often have.

CHAPTER 13

Our Responsibility as Creators

We are creators. We come from a divine lineage. We are all created in the image of God. We are thought forms created by God. We are given free will. We can choose to live in Harmony, Balance, and Love. Or we can choose to live a life taken up with grabbing for power and material wealth. It is up to us to create the world we want in the image of our mind, our best thinking.

We *are* our thoughts, and our creations are based *on* our thoughts. If we choose to live a life in fear, anger, and hatred, that is the world we get to live in. If we choose Love, Truth and Harmony, *that* is the world we get to live in. If we understand that we are all One, all the children of God, and that there is really no separation among all of us, then we may behave in a balanced and harmonious manner.

In the previous chapters, I have given evidence that you are much more than the physical body. God has given all of us the same tools, which He utilizes. We have consciousness, we are aware, we have free choice, and we have the ability to create with our Thoughts and Words. But we also are given a dense body, where it is a bit harder to create than in the spirit world.

However, we can still create—and we can know and realize the vibrational level of our essence, the I AM. Our dense bodies and our forgetting who we really are need not discourage us: it gives us the chance to learn.

This chapter is my interpretation of why we are here, and some of the key lessons we are here to learn.

Our World Is Ours to Create

Our wars, our hatred of others who do not look like us, our grabs for power and wealth, our disregard for other life forms, whether plants, animals, inhabitants of the sea and air—all these manifestations of destruction are the results of our own creation. God did not create this. God just established the *rules of creation* through the spiraling lights to create matter and life. He has given us free will to create the world we want to live in. He has tried to help us remember our heritage through avatars He has sent to us (such as Christ, Buddha, Krishna, Mohammed, and others). We have the choice to create whatever world we want.

We are in an age where we are starting to remember our divine heritage. My purpose in writing this book is to help you look behind the obvious physical aspects of yourself and to see your true nature. There are reminders of your true nature because of what you can do with intention and thoughts. In this book, you have been given a tool to see into the invisible world of subtle energies, the energies behind the physical world.

There are layers above the physical world described in ancient texts (e.g., the Vedas of India, the teachings of our Avatars). In BioGeometry, there are the layers above the physical—the vital, emotional, mental, and spiritual. These are real worlds, where we can find the basis of problems that affect us in the physical world. Quantum physicist David Bohm described some of these worlds as the *implicate realm*. It is from the implicate realm that the physical, *explicate* world is derived.

I have also been given the privilege of a glimpse into the world of the afterlife through the eyes and ears of my good friends Jeanne Love and Regina Ochoa. I have had conversations with the deceased astronauts from the Challenger Space Shuttle and listened to the Columbia Space Shuttle astronauts describe their fiery deaths and an amazing glimpse into the worlds they now see and experience (see *www.challengercc.org* and *www.fmbr.org*). In fact, I offered to transcribe the spoken words of deceased Columbia astronaut Sally Ride, who was one of the hosts for Regina and Jeanne's channeling session. Since I was working so very hard at this transcription, I started to channel her words and voice—her gift to me, so I could have a channeling experience.

The deceased have a much closer view of what creation is about. They create their image, and the clothes they wear. They create their living habitats. They understand that all time is now. They help relatives and friends cross the astral moat and live a new life. They learn in the afterlife some of the information that I am now communicating to you. They often state that they wish they had known more about the afterlife while they were alive. Then, they would have lived their lives differently.

When Creating, Think "Love, Harmony, and Balance"

My key message to you is that YOU are the creator of your life and the nature of the world in which you (and all of us) live. We are all One, but we need to understand this at a deeper level so that we do not do the harm we sometimes have done to one another.

Knowing all this, I try to think in terms of what *I* would do when creating my world. Since we are all given free will, we cannot dictate what others will do. We must understand that free will results in individual decisions, some of which we will not agree with—because, like everyone else, we are individuals who have had different experiences, which will create different ways of thinking. Thus, when dealing with others, we need to listen, hear the various opinions, and try to reach a point of balance. This may not satisfy everyone, but it will be harmonious enough so that all parties will get enough of what they want, so that there is a balance—without the need for war or killing. And since we are One, we need to learn forgiveness and being non-judgmental.

Thus, when I am in Creator Mode, my main focus is Love, Harmony, and Balance. I cannot impose my will on others. Everyone must come to a sense of balance from their own choice. I believe God has done this when He gave us free will so that we could become individuals. However, to avoid doing harm to ourselves and others, He has allowed us to grow and learn. If our goal is to always create in Love, Harmony and Balance, then we will create our own Golden Age.

This is what's behind Joachim Wippich's way of thinking. He never *asks* anyone to do anything, because asking is akin to a command. Instead, he *invites*. If he is asking a person to heal himself or herself, he has them invite the cells or organs to come into harmony with the disharmony. He knows that the cells will then know that whatever they were doing is causing disharmony, and therefore they will do whatever is needed to come into harmony. This is how healing occurs. Joachim also reminds people of their I AM essence. Once they remember, their vibrational level is increased to their I AM, and the inner peace and healing occurs automatically.

My summary statement for this chapter—and indeed, for the whole of this book—is quite simple:

When creating, think Love, Harmony, and Balance.

PART VII

Bonus Section

The Wisdom of Joachim Wippich

Update on Affirmations by Joachim Wippich

The Wisdom of Joachim Wippich

Joachim simplifies his key teachings within the words "Harmony" and "I AM." Within these words are a wealth of wisdom and understanding of enlightenment. If humanity understood these few words and lived by those principles, the world would be a place of joy, peace, and happiness.

Within the word "Harmony," everyone would understand that in all situations—since we are given Free Will and we are inherently composed of differing polarities of light—there will be differences that may cause disharmony. But with the understanding of harmony, a balance will be agreed upon and there is no need for the wars and fights that often occur in our world today. Our body will also understand harmony with possible disharmonies, so that when the I AM consciousness invites our cells to come into harmony with their disharmony, healing will occur.

Within the words "I AM" is the meaning of Everything:

- There is only One, and each of us is the individuation of the One. Everything and everybody is part of the One.
- "I AM" is a statement that we are One with God.
- We are all created in the image of God. All is open to us if we can truly open up to the I AM.
- We are all Creators.
- We are given forgetfulness of who we are, and we live in a density level that does not make it easy to know that we are Creators. This situation enables us to live, learn, and evolve.
- Evolving in a dense level allows us to learn to create in a responsible manner. It allows us to learn the meaning of Love, Harmony, and Balance, and thus to create accordingly.

- God has given us Free Will to experience whatever we want to experience. In so doing, God also experiences Everything.
- As we each seek greater wisdom, understanding, and knowingness in a state of Love and Harmony, our beingness becomes opened to a deeper understanding of who we are, the I AM.
- At the I AM vibrational frequency, we have better communication with our cells and tissues so that healing occurs. True healing comes from within.
- There are levels to the understanding of I AM.
  - » For an adept starting out, I AM may mean that we are something more than the I am ego personality.
  - » I AM may then evolve to the existence of God and that we are created by God.
  - » Further evolution may translate to: God is in everything and that is how everything is made.
  - » A further evolution may translate to: we are God and everyone is God and we are all One.
  - » A further evolution is that everything is Thought.
- In healing, inviting your cells to come into harmony with their disharmony, and coupling that with an affirmation to bring yourself to your I AM vibrational level, often results in healing.

There are many corollaries associated with the I AM concept:

- Let's start with the previous statement: Everything is thought. Our reality is created by our thoughts. When we are angry or fearful, it is our thoughts. When we react to a situation, it is our thoughts.
- When we hurt someone by words or action, it is our thoughts. Then the law of karma comes in so that lessons can be learned in the school-of-life experiences.
- When we state "I AM thoughts within thoughts within thoughts" and say this with 3 "thoughts," then 6 "thoughts," and then 9 "thoughts," we move to a high vibrational state

(since we are then moving to the initial state of our thought creation when we knew our true identity).

- When we state "I AM Everything," this acknowledges who we really are, and our vibrational level goes up.

Our choice of Words and Thoughts are very important. In school, we are often taught that sticks and stones can hurt but that you cannot be hurt by words. That is incorrect, since we are creators and we create with every word and thought. We must be aware of the impact of our words and thoughts, even in the absence of the person we are directing it toward, since it has an impact. We saw that when we carried out experiments and asked subtle energy fields to move and they moved. We asked for harmonization and harmonization occurred.

Let's look at some key words in Joachim's world.

Rethinking the Rethink Concept

In many of Joachim's affirmations, he uses the word "rethink." A primary example is: "I AM rethinking, rethink I AM."

You might ask, "Why 'rethink'? I know who I am." The problem is that our senses and ego lead us to assume that we know the answer—the truth. Our teachers, parents, society, and priests all give us perspectives on answers to questions we might have. We thus form our perspectives and we call that the "truth," or correct answer, to whatever our question is.

However, if we start to think about most questions and statements, we realize that different people have different answers. To find the best answer or validity to a statement, it is best to "rethink" the statement or question.

Finding the answer to the truth of questions or understanding deep statements such as the meaning of "I AM" requires rethinking. Answers to such questions are often not straightforward. In fact, you can go down any one of a number of lines of thinking. Just as creation of matter has light following a spiraling path, so finding the truth leads us down a spiraling path. We need to allow the spiraling "rethink" path to spiral us to the core, where there is realization that there is only one answer, and that answer is I AM. Then all the other statements and questions make sense. We realize that the validity of the many parts of the "I AM the I AM" poem by Francis Vaughan is true (see this poem in *The Seeker and The Teacher of Light*). We realize the truth of the statement, "I AM Everything."

As a general philosophical point, when seeking the truth of anything, one should follow a spiral path of thinking. A spiral means that you are rethinking all aspects of the question. By doing so, you are spiraling into an answer in which you have given consideration to all aspects, so when the truth is known, you know it in your heart and mind. Thus, when you say "I AM rethinking" (pause) "rethink I AM," you allow yourself to move closer to the truth. And your vibrational level goes up.

Another Way of Looking at Thoughts and the Written Language

In this section, we have been exploring the power of thoughts. Thoughts are the basis of everything. Joachim's Affirmations are thoughts with words carefully chosen to carry the meaning of the thoughts. Sometimes the flow of words are not in the best English, and there is no punctuation. His students have sometimes tried to change the words to express better English and have put in punctuations. These efforts fail and the resulting affirmations do not have the power/vibrational frequency to bring a person into harmony or to the I AM frequency.

In Chapter 5 of this book, "Subtle Energies—A Developing Story," one of the key points made by Olga Strashun was that thoughts are powerful subtle energies. I then started thinking about the nature of thoughts and their flow. Thoughts do not have punctuation marks. They have a flow of words conveying our thoughts.

I decided to carry out an experiment. I took the previous section on "Rethinking the Rethink Concept" that you just read, and I took out all punctuation marks; then I tested that section to see if it had any BG3 or 3-6-9 subtle energy qualities. It had none—just like the original words with punctuation. The section still had the appearance of the original section and did not seem to flow as thoughts.

I then laid out the words to more closely resemble the flow of thoughts. When I did this, I found that the section now had a BG3 and 3-6-9 subtle energy level of 300. I have repeated the "Rethinking the

Rethink Concept" below as an unpunctuated flow of thoughts, so that you can do your own testing with either your BG3 or 3-6-9 pendulum.

Rethinking the Rethink Concept as Thoughts

Many of Joachim's affirmations use the word rethink

A primary example is I AM rethinking rethink I AM

You might ask why rethink   I know who I am

The problem is that our senses and ego lead us to assume that we know the answer the truth

Our teachers parents society and priests all give us perspectives as to answers to questions we might have

We form our perspectives and we call that the truth or correct answer to whatever our question is

If we start to think about most questions and statements we realize different people have different answers

To find the best answer or validity to a statement it is best to rethink the statement or question

Finding the answer to the truth of questions or understanding to deep statements such as the meaning of I AM requires rethinking

Answers to such questions are often not straightforward

In fact you can go down any one of a number of lines of thinking

Just like creation of matter has light following a spiraling path finding the truth leads us down a spiraling path

We need to allow the spiraling rethink path to spiral us to the core where there is realization that there is only one answer and that answer is I AM

Then all the other statements and questions make sense

We realize the validity of the many parts of the I AM the I AM poem by Francis Vaughan is true

see poem in The Seeker and The Teacher of Light

We realize the truth of the statement I AM Everything

When seeking the truth of anything one should follow a spiral path of thinking

A spiral means you are rethinking all aspects of the question

By doing so you are spiraling into an answer in which you have given consideration to all aspects

so when the truth is known you know it in your heart and mind

Thus when you say I AM rethinking rethink I AM you allow yourself to move closer to the truth

And your vibrational level goes up

The concepts presented in this section should give us pause to consider what are the best ways to express our thoughts. Should our textbooks be constructed in a way to maximize the flow of thoughts, rather than blocks of information to be memorized or learned?

Update on Affirmations by Joachim Wippich

Joachim Wippich is continuously evolving his affirmations for both healing and increasing our vibrational level to be at the I AM level. This chapter presents an updated, evolved version of the affirmations presented in *The Seeker and The Teacher of Light*. (Note that IFSG and BG-EHS Practitioner Jan Walsh will be making periodic updates of affirmations on an ongoing basis on my website. *See https://jerrygin. com/affirmations/*.)

The Updates are divided into two sections.

The first is called "Securing Your Resonance." Its purpose is to bring about a state of harmonization between the two parts of the brain and the heart. Such harmonization brings about a better flow of energies in the body, and is a good precursor for both dowsing and radiesthesia, as well as for stating the affirmations intended to bring you to the I AM vibrational level. In addition, if there are energetic interferences on your property, the affirmations in this section will correct them.

This is followed by a discussion of ways I have found to be effective in stating affirmations to reach the I AM vibrational level.

The second section has been put together by Jan Walsh. Jan works closely with Joachim on updates of the affirmations. (As mentioned above, Jan has, and will continue to, put those affirmations on my website: *https://jerrygin.com/affirmations*.) The section has a list of the various affirmations.

PART 1.
Securing Your Resonance

To be sure that you are in optimal resonance with your personal energies/vibrations—especially for dowsing and determining your vibrational level—there are three steps to follow:

1. Bring your brain and heart into harmony (*see* "Harmonization: Right Brain / Left Brain Exercise," below).

2. Bring yourself to the I AM vibrational level.

3. Harmonize any possible environmental factors that might affect you and your dowsing.

The following steps show how Joachim Wippich secures a personal resonance before dowsing. Taking into consideration how the environment may be influencing your dowsing is also an important element that often is overlooked.

1. Harmonization: Right Brain / Left Brain Exercise

Hold your hands out to the side and then bring them together in front of your heart, while saying each of the following lines:

Women:

"I AM inviting Every thought within my left brain and my right brain to join together with my heart I AM"

Men:

"I AM inviting Every thought within my right brain and my left brain to join together with my heart I AM"

This exercise can be done any time you think you might be out of balance, not just before dowsing. It is also a beneficial practice to begin every morning.

2. I AM Affirmation: Vibration Level—"I AM Everything I AM"

Make this statement—"I AM Everything I AM"—and check for "Yes" or "No" using your dowsing tool. Even if your dowsing tool indicates "Yes," check what percent is "Yes" by stating the following:

"My I AM is more than 10%."

If "Yes," continue.

"My I AM is more than 20%."

And so on, until you reach 100%.

If your dowsing tool gives a "No" at a percentage lower than 100, say:

"I AM Rethinking Rethink I AM"

(and then pause)

"I AM Rethinking Rethink I AM"

(pause)

"I AM Rethinking Rethink I AM"

(pause).

The reason you pause between repetitions is to absorb what you are saying.

Now recheck to see what percent you are. Repeat the above process until your "I AM" resonance reaches 100%.

3. Harmonization of Location

We live in an environment, and it can influence the accuracy of our dowsing. The following steps will help to neutralize its influence.

"I AM in Harmony with My Location"

"My location is in Harmony with Me"

After making this statement, check for "Yes" or "No" using your dowsing tool. If you get a "No" to either line, say the following correcting statements:

"I Neutralize Normalize Harmonize Energize Polarize
this property"

"It will be beneficial for my entire I AM being"

Recheck the initial statement that came out "No," and if necessary repeat the process of saying the correcting statements until your dowsing tool gives you a "Yes."

The Wisdom of the Many I AM Affirmations

There are a number of Affirmations that will bring you to the I AM vibrational level. They will all work. It just depends on which affirmation resonates with you most to bring you to the I AM realization and vibrational state.

Rethink

Many of the affirmations start with "I AM Rethinking Rethink I AM." As described in the chapter on "The Wisdom of Joachim Wippich," *rethink* is a powerful word: it activates your inner knowing towards the understanding that you are I AM. I view this as "spiral thinking"—that is, your thinking spirals to the center, where there is full knowledge of who you are, I AM.

Thoughts

Another spiral involves thoughts. You are Thoughts, and Everything is Thoughts. Thoughts are also your I AM essence. Thus, the affirmations that state "I AM Thoughts I AM" or "I AM Thoughts within Thoughts within Thoughts" repeated 3, 6, and 9 times bring you into the I AM state—that is, back to your original evolutionary thought creation—and you remember who you are.

You can also introduce the rethinking concept with the thoughts by stating "I AM rethinking thoughts within thoughts within thoughts I AM." It will lead you to the same place—I AM.

Another train of thought that leads you to the I AM is the statement, "I AM Everything I AM." Realizing that you are Everything brings you to the I AM vibrational level. If that statement feels like too much, then other affirmations may work better for you. We are all evolving in our understanding of I AM.

Self-Love

People are not taught that every person is amazing and of divine lineage. Power is given to others. People sometimes forget about self-love. For some, it is difficult to state, "I Love myself." Thus, in bringing about the remembrance of who we are, Joachim will sometimes use the terms "I AM Amazingly Delightful" or "I AM an Amazingly Delightful Treasure." This helps bring the person into the I AM vibrational level.

Harmony

I AM Affirmations often will use words to bring in Harmony. There may be disharmony at various levels: physical, etheric, emotional, mental, spiritual. Thus, there are statements to bring in Harmony with the disharmony at various levels.

I am giving this introduction to the I AM affirmations since you may wonder, "Why so many affirmations?" It is because different people will have different understandings, and Joachim has created affirmations to take these differences into account. The affirmations in Part 2 reflect this thinking.

The Letters of "I AM"

The letters of I AM are sometimes defined by Joachim as follows:

The "I" stands for intelligence, the "A" is acknowledgement of I AM, and the "M" is for memory of who you are. The "AM" also indicates continuity.

Finding the Affirmation That Works Best for You

Try the various affirmations and see which one works best for you. You will know this as you observe your vibrational level, either by dowsing or by using your personal wavelength, and watching the pendulum or L-rod rotate faster and faster. When you are fully in the I AM state, there is centering stillness and the rotating pendulum becomes still.

Any affirmation can be repeated 3 times, then 6 times, and then 9 times. The power of the affirmations goes up with each repetition, as evidenced by the rate of rotation of your pendulum.

Affirmations and Problem Solving

Many of the affirmations stated in this section are also embedded in affirmations for other purposes. That is because reaching I AM is key for working on many problem areas.

Completing the Affirmation

All the affirmations below should terminate with "I AM (full birth name) I AM."

Coordinating I AM with Breath

You can also incorporate your breath in your "I AM" affirmations: As you inhale, think "I"; as you exhale, think "AM." This adds to the effectiveness of the I AM.

The Concept of Evolving

In various affirmations, Joachim uses the term "Evolving." If you are evolving, it means you are making progress in your life's journey (or soul's journey). Joachim uses this term in a variety of applications. As an example, you may be wondering if you should purchase a particular book or if you should take a self-help seminar. You would then dowse to see if buying the book or taking the seminar will result in your evolution in this lifetime.

The Home Concept

Joachim teaches that you should be "home" most of the time. That is, you should be experiencing and being in the "now and here." If your mind is elsewhere, such as being preoccupied with elements of the past (for example, various problems you have had), then you are not "home." If you are always projecting into the future and not living in the "now," that can also be a problem. For emphasis, Joachim says that you should not be away from home more than a few minutes every 2 hours.

The Stealth Body Concept

In some of Joachim's affirmations, he refers to the "stealth body." This refers to the "unseen energies" that can sometimes affect us.

Harmony Spirits

When an affirmation talks about "Harmony Spirits," Joachim is referring to one's guides or angels.

Use of Punctuation Marks

It should be noted that there are no punctuation marks in the descriptions of the various Affirmations in Part 2. Working with Joachim and Jan, it was discovered that punctuations interfered with the flow of thoughts. Joachim always knew that punctuation marks interfered with Affirmations. They also turn out to interfere in the descriptions of Affirmations, as evidenced by a significant drop in the BG3 and 3-6-9 subtle energy levels of the whole section of Affirmations below when marks were put into this section.

This discovery is a huge paradigm shift on how our mind (consciousness) works. We are all taught grammar and punctuation. But is that truly how our minds work? When we think, it is always a flow. There may be pauses, but there is a flow to achieve completion of the thoughts you are trying to present. It may be that the best way to present thoughts is without punctuation, and to just use space and paragraphs in expressing flow.

PART 2.
Joachim Wippich Affirmations
as Interpreted by Jan Walsh

This document is an attempt to explain and share some of the remarkable affirmations that I have learned from my friend and teacher Joachim Wippich   Rather than focusing on the correctness of or agreement with the information I urge you to experience the measurable energy frequency of the document in its entirety   The stillness it generates is palpable   By removing some (but not all) of the punctuation Joachim discovered that energy flow and harmony increased

List of Affirmations

I AM
  Evolving

Thoughts
  Intefering Thoughts
  Blockages

Knowledge
  Affirmations Related
  to Knowledge
  Protection
  Harmony Affirmations
  Meditation
  Procrastination
  Memory
  Lost Item

Helping the Planet

Health
  Property
  Physical Problem

Pain or Discomfort
Cells
Immune System
Healing Affirmations
Yawning
Depression
Stress
Age
Death
Weight

Relationships
  Relationship Situation
  Family
  Helping People

Pets

Travel

Words

Blessings

I AM

"I AM" is the pure loving essence of everyone

The following affirmations are helpful in bringing your "I AM" to 100% so that no interfering Thoughts can come in and you are in a state of balance and harmony   As you refer to yourself that way you open up your heart and mind for Divine Inspiration Knowledge (For a complete explanation of "I AM" refer to Chapter 6 in *The Seeker and The Teacher of Light* by Jerry Gin)

"I AM Everything I AM"

"I AM Thoughts I AM"

"I AM owning my own Thoughts I AM"

"I AM responsible for my own Thoughts I AM"

"I AM in Harmony with my Disharmony I AM"

"I AM Self-Correcting my Disharmony Thought Creations I AM"

"I AM fulfilling my evolutionary self-correcting Thought creations I AM (full birth name) I AM 24/7"

"I AM the most Amazingly Delightful Precious Treasure in this Life Experience within the Universe I AM"

(It is very important that you *believe* this)

"I AM the most Amazingly Delightful Thoughts Creation I AM"

(For a complete explanation of "I AM" refer to Chapter 6 in *The Seeker and The Teacher of Light* by Jerry Gin)

Try each affirmation and see which works best to raise the "I AM" to 100%   Remember to check the efficiency and effectiveness of the affirmation by dowsing each component of "I AM" for percentage   The "I" stands for intelligence the "A" is acknowledgment and the "M" is for memory   We are energy intelligence which is the "I"   Acknowledging we are energy intelligence is the "A"   Remembering who we are as energy intelligence is the "M"

After saying an affirmation dowse the following

"My 'I' is more than 10%, more than 20%" etc. (Dowse Y/N)

"My 'A' is more than 10%, more than 20%" etc. (Dowse Y/N)

"My 'M' is more than 10%, more than 20%" etc. (Dowse Y/N)

This number is a reference point to help get to 100% with each component of "I AM"   If the number is less than 100% repeat "I AM Rethinking Rethink I AM" 3 times   Check the reference point again to see if it has moved higher   Continue this process until 100% is reached on each of the 3 letters

One affirmation may always help to reach 100% with all 3 components ("I" "A" "M")   Consider this as a "go to" affirmation throughout the day ensuring I AM "home" with no voids for interfering Thoughts to come in   Repeat it often   The affirmations are more powerful when repeated 3 times and pause 6 times and pause then 9 times

(Refer to *The Seeker and The Teacher of Light* for information on 3-6-9 energies)

"I AM inviting my Thoughts to come home and rethink with me
100% 24/7"

This affirmation is another way to bring myself to 100% "I AM" Inviting allows free will   "I AM" has several levels   There is a physical etheric emotional mental spiritual and stealth (unseen) level   Each can be checked separately to see if it is 100% by saying the following

"I AM in Harmony with my physical body I AM"

(Dowse "Yes"/"No" Check percentage)

"I AM in Harmony with my etheric body I AM"

(Dowse"Yes"/"No" Check percentage)

"I AM in Harmony with my emotional body I AM"

(Dowse "Yes"/"No" Check percentage)

"I AM in Harmony with my mental body I AM"

(Dowse "Yes"/"No" Check percentage)

"I AM in Harmony with my spiritual body I AM"

(Dowse "Yes"/"No" Check percentage)

"I AM in Harmony with my stealth body I AM"

(Dowse "Yes"/"No" Check percentage)

To increase the percentage repeat "I AM Rethinking Rethink I AM" 3 times

The efficiency and effectiveness of each level can be checked as well

The breath can also be incorporated in "I AM" affirmations
As I inhale I rethink "I" as I exhale I rethink "AM"
Do this several times to increase the "I AM" presence

(For more breath explanations look at page 141 in Jerry's book *The Seeker and The Teacher of Light*)

If someone struggles to understand "I AM," they should say the following once a day

"I AM Rethinking Rethink I AM"

"I AM Rethinking Rethink I AM"

"I AM Rethinking Rethink I AM"

(Refer to *The Seeker and The Teacher of Light"* page 141 for an explanation of Rethinking and to the previous chapter)

Evolving

When repeating different affirmations I can dowse to see if I AM evolving or not then check the vibrational level   When selecting an activity it is possible to dowse to see if I AM evolving by participating in that activity   (See the Knowledge section for more information)

Remember when dowsing to make a statement instead of asking a question for more accuracy and Secure Your Resonance first (Check the Dowsing section for the steps to Secure Your Resonance)

"I AM owning my Evolutionary Thought Creation 100% I AM"

Checking the evolutionary number is another way to see if I AM Evolving To understand these affirmations the evolutionary number needs to be 350 or more according to dowser Bob Mahany

"My evolutionary number is more than 100"

If the dowsing tool indicates "Yes" to the above statement keep increasing the number by increments of 100 until reaching "No" This is a reference point You can increase the evolutionary number by repeating any of the following Affirmations   Repeat the affirmation 3 times and then check to see if the evolutionary number has increased

"I AM Rethinking Rethink I AM"

"I AM Thoughts I AM"

"I AM Everything I AM"

"I AM in Harmony with my Limitations I AM"

"I AM my Limitations I AM"

"I AM Evolving my original Thought Creation I AM"

"I AM Evolving my Universal Law Thought Creation I AM"

"I AM Evolving my own Thoughts I AM"

"I AM owning my Evolution I AM"

"I AM fulfilling my Divine Christ God Collective Consciousness Thought Creation without Limitation Thoughts I AM (full birth name) I AM"

(For an explanation of the use of "full birth name" refer to *The Seeker and The Teacher of Light* page 122 References to Divine are found on page 118 and information on your vibrational level is located on pages 86-90)

A name has a vibrational frequency associated with it If I have been known by several names I can say each one every once in awhile and repeat the following

"I AM Rethinking Rethink I AM"

(3 times and pause 6 times and pause then 9 times) to harmonize the name

Names other than my full birth name can carry controlling frequencies with them A nickname I have may have an attachment with it from the person who gave it to me

"I AM in harmony with my physical name I AM" (Dowse Y/N)

"I AM (full birth name) amazing I AM" (Dowse Y/N)

"I AM the most amazingly delightful I AM being I AM"
(Dowse Y/N)

Thoughts

Thoughts are information   Thoughts are Everything
Everything is a Thought

The following affirmations help to bring your "I AM" into harmony using your Thoughts

"I AM inviting myself to come to the correct vibrational frequency of my first Thought Creation I AM"

"I AM inviting my Thoughts to come home and be with me 100% 24/7 I AM"

"I AM inviting my Thoughts to self-correct I AM"

"I AM My Thoughts I AM"

"I AM Thoughts I AM"

"I AM Consciousness I AM"

"I AM inviting Thoughts to Rethink with me I AM"

"I AM Rethinking Rethink I AM"

"I AM my Evolutionary Divine Thought Creations 24/7 I AM"

"I AM accepting responsibility for my Thoughts I AM"

"I AM Every Thought I AM (full birth name) 24/7"

(If everyone Rethinks this, the world will experience peace)
(Chapter 5 of *The Seeker and The Teacher of Light* by Jerry Gin covers how thoughts influence us)

Dowse percentage to see if 100% of Thoughts are being used

Example: "I AM using more than 50% of my Thoughts I AM" (Dowse "Yes" or "No") If the percentage is less than 100 say the following and then recheck the number

"I AM Rethinking Rethink I AM"

"I AM Rethinking Rethink I AM"

"I AM Rethinking Rethink I AM"

Depression anger and sadness can be signs of interfering Thoughts When I AM not 100% "I AM" interfering frequencies can affect me   If I AM looking at an old picture of myself and seem unhappy I can dowse to see if my Thoughts were home 100%

The following is a powerful affirmation that can create stillness

"I AM inviting Thoughts within Thoughts within Thoughts to rethink with me"

(Pause)

"I AM inviting Thoughts within Thoughts within Thoughts within Thoughts within Thoughts within Thoughts to rethink with me"

(Pause)

"I AM inviting Thoughts within Thoughts within Thoughts within Thoughts within Thoughts within Thoughts within Thoughts within Thoughts within Thoughts to rethink with me"

(Pause)

"I AM Entertaining Maintaining Sustaining and Supporting Divine Love Divine Harmony Divine Gratitude Divine Forgiveness Divine Light Divine Life Divine Happiness Divine Joy Divine Wisdom Divine Oneness Divine Tranquility Divine Humility Divine Inner Peace I AM Rethinking Rethink I AM (full birth name) I AM"

What we do to others, we do to ourselves ten times over Instead of blaming others apologize because they are not awake and don't know

There is no false ego only incorrect ego because it is man made

Our incorrect ego is a part of us Thus it is good to acknowledge our incorrect ego and be in harmony with it

"I AM in Harmony with my incorrect Ego I AM"

"I AM my incorrect Ego I AM"

Interfering Thoughts

If I measure owning my own thoughts at less than 100% there is an interfering Thought present   Some may refer to this as an entity whose Thought can control the physical etheric emotional mental spiritual and stealth body making me feel drained   An entity (Thought) contacts me because it wants my help as an intelligent consciousness   I can dowse to determine the source of the interference but that is not necessary to correct it   For more knowledge of the situation dowse

"I AM aware of you but don't know you"

If "Yes" it is an unrecognized frequency   If "No" dowse to see if it is friend relative acquaintance or ancestor   For either "Yes" or "No" dowse the following to determine how quickly a response is needed

"It requests my help"

(urgent need) (Y/N)

"It needs my help"

(you have time to respond) (Y/N)

"It seeks my help"

(lowest level of priority) (Y/N)

I don't get rid of an entity (Thought) I invite it to come into harmony I invite the energy to experience the harmony affirmations

"I AM inviting you to come to the correct vibrational frequency of your first Thought Creation"

"I AM inviting you to come to the correct vibrational frequency to correct the Karmic situation"

"I AM gifting you the Harmony Affirmation"

"I AM inviting you to Rethink the Harmony Affirmation"

"I AM Rethinking Rethink I AM"

The best way to protect yourself is to become familiar with Securing Your Resonance (Part 1 of this chapter) checking it often   Know that when I AM upset or angry or sad I have allowed an Interfering Thought to come in

This happens because the "I AM" is not 100%

Telepathically invite someone who is mean to Rethink with you Love and Understanding   The *person* is not bad there is a *Thought* influencing them

Blockages

Blocks are considered controlling frequencies

Blockages may be caused by my own Thoughts or someone else's If I AM too generous I allow others to think for me   Over-Generosity occurs when the mind is not presently active within the "I AM" on every level of "I AM" (Part 1 "I AM Affirmation")   At this point disharmony can enter and cause suffering   Dowse to see if there are any blockages and how many

"I have blockages"

(Dowse "Y/N and how many)

To correct a blockage, state the following

"I AM 100% (full birth name) I AM"

"I AM owning my own thoughts I AM"

"I AM rethinking rethink I AM"

(say 3 times and then pause)

"I AM rethinking rethink I AM"

(say 6 times and then pause)

"I AM rethinking rethink I AM"

(say 9 times)

"I Neutralize Normalize Harmonize Energize Polarize to what is beneficial for my entire I AM being"

(To understand the reason for saying something 3, 6, and 9 times refer to page 23 of *The Seeker and The Teacher of Light*   You will also find an explanation for "rethinking rethink" in Part 1 of this chapter.)

The following affirmations are helpful in preventing blockages

"I AM fulfilling my evolutionary self-correcting Thought Creations I AM (full birth name) I AM 24/7"

"I AM in harmony with my limitations I AM"

"I AM my limitations I AM"

"I AM Rethinking my limiting Thoughts I AM"

"I AM (full birth name) I AM"

Knowledge

When your thoughts are 100% then knowledge can come in

If I read a book take a class go to a movie or watch TV etc. I can check to see if it will benefit my "I AM" being   I should trust myself when I question something because there could be a message   First I Secure my Resonance (Part 1) then dowse to see if I AM evolving when participating in the activity

"I AM evolving 100% I AM"

(check your dowsing tool for a "Yes or "No" response)

These experiences may have a controlling vibrational frequency Don't allow it to come in and control   I need to be cautious of exposing myself to interfering frequencies before participating in an activity and I do so by saying

"I Neutralize Normalize Harmonize Energize Polarize [the activity] so that it is beneficial for my entire I AM being"

After participating in an activity I can bring myself into Harmony by saying

"I AM 100% I AM"

"I AM the balance between my Thoughts and Knowledge 100% I AM"

"I AM Rethinking Rethink I AM"

Dowsing the following statements is another way to check the "I AM" level

"I AM allowing brainwashing in my life I AM"

(Measure percentage if "Yes")

"I AM self-brainwashing my left and right brain I AM"

(Measure percentage if "Yes," checking each side)

If "Yes" is indicated from your dowsing tool, then state the following to correct the blockage

"I AM 100% (full birth name) I AM"

"I AM owning my own thoughts I AM"

"I AM Rethinking Rethink I AM"

(say 3 times and then pause)

"I AM Rethinking Rethink I AM"

(say 6 times and then pause)

"I AM Rethinking Rethink I AM"

(say 9 times)

(To understand the reason for saying something 3, 6, and 9 times refer to page 23 of Jerry Gin's book *The Seeker and The Teacher of Light* You will also find an explanation for "rethinking rethink" on page 122 of that book)

Affirmations Related to Knowledge

"I AM knowledge I AM"

"I AM using my Thoughts in a different way I AM"

"I AM (full birth name) I AM"

"I AM aligning myself with Universal Law I AM"

"I AM in balance with Universal Intelligence I AM"

"I AM aligned with God Thought Creation I AM"

"I AM in Harmony with Universal Law I AM"

(use if your mind is wandering)

"I AM opening my door of knowingness"

"I AM inviting greater Knowledge to come through and creative Thoughts I AM"

"Mind go to the Universal Mind and bring back what I need to know for the next step in my Divine Evolutionary Journey"

"Mind do not be gone more than 2-3 minutes every hour for no more than 2 hours"

Protection

Fear is false evidence appearing real When all your Thoughts are "home" there is no room for other Thoughts

"I AM using 100% of my Thoughts I AM"

(Repeat this 3 times to bring yourself to 100% "I AM")

Before going to bed going to a gathering or party I check that I own my own Thoughts and that I control my own Thoughts

"I AM inviting myself to give myself permission to come into Harmony with my own Disharmony I AM"

By coming into harmony with disharmony I disempower any non-beneficial energy "It no longer needs to be neutralized released cleared transformed or shooed away" (Jerry Gin)

If you are amazing, then no one can break into your field

"I AM (full birth name) amazing I AM"

"I AM the most amazingly delightful I AM being I AM"

When a bubble is created around a person for protection it is challenging others to break it   Instead say the following

"I AM surrounded by Divine Light so that no negativity will reach me or become me I will accept that which is beneficial for mankind I AM"

When I get upset I am not 100% I am in disharmony   What part of my "I AM" am I not controlling (Physical etheric emotional mental spiritual stealth)

Dowse and correct each level if "No"

"My own Thoughts are in my Physical body 100%"

"My own Thoughts are in my Etheric body 100%"

"My own Thoughts are in my Emotional body 100%"

"My own Thoughts are in my Mental body 100%"

"My own Thoughts are in my Spiritual body 100%"

"My own thoughts are in my Stealth body 100%"

Dowse and correct each if "No"

"I AM in Harmony with my Physical Thoughts I AM"

"I AM in Harmony with my Etheric Thoughts I AM"

"I AM in Harmony with my Emotional Thoughts I AM"

"I AM in Harmony with my Mental Thoughts I AM"

"I AM in Harmony with my Spiritual Thoughts I AM"

"I AM in Harmony with my Stealth Thoughts I AM"

To correct, say

"I AM Rethinking Rethink I AM"

"I AM Rethinking Rethink I AM"

"I AM Rethinking Rethink I AM"

Harmony Affirmations

By Harmonizing you disempower any non-beneficial energy Thoughts

Love is involved in the Harmony Vibrational Frequency

After repeating any of these Harmony Affirmations check the vibrational level or evolutionary number as explained in the I AM section of Affirmations

"I AM the balance between my harmony and disharmony I AM"

"I AM in Harmony with my Disharmony 100% I AM"

"I AM in Harmony with my own I AM being I AM"

"I AM owning my entire I AM being I AM"

"I AM self-correcting my disharmony Thought creation I AM"

"I AM accepting myself I AM no matter what circumstances present themselves to me I AM"

"I AM Rethinking Rethink I AM"

"I AM Rethinking Rethink I AM"

"I AM Rethinking Rethink I AM"

"I AM fulfilling my evolutionary Divine Christ God Collective Consciousness Thought Creation I AM (full birth name) I AM"

"I AM inviting every Thought Creation to please rethink with me 100% I AM 24/7 (full birth name) I AM

Divine Christ God collective consciousness Thought Creations I AM thanking for every blessing of every Thought Creation on this

wonderful planet and the Universes throughout my evolutionary
Thought Creation I AM (full birth name) I AM"

(Rethink this 3 times, 6 times, and 9 times for the most effect)

"I AM responsible for my knowledge Thought Creations I AM"

"I AM 100% I AM"

"I hope this will help Every Thought Creation I AM"

Meditation

It is difficult for anyone to meditate at the correct harmonic vibrational frequency

When you meditate and go to another place, you don't know in advance whether it is beneficial or not Make sure you dowse to see if you are evolving before and/or after experiencing a meditation If I AM not "home" my "I AM" is not 100%   You can say the following

"I AM Rethinking Rethink I AM"

"I AM Rethinking Rethink I AM"

"I AM Rethinking Rethink I AM"

To create the centered stillness that may be found when meditating try repeating the following phrases to achieve that stillness more quickly

I AM inviting every Thought around me to come into harmony
with its disharmony

"I AM inviting Thoughts within Thoughts within Thoughts to
Rethink with me"

"I AM inviting Thoughts within Thoughts within Thoughts within
Thoughts within Thoughts within Thoughts to Rethink with me"

"I AM inviting Thoughts within Thoughts within Thoughts within
Thoughts within Thoughts within Thoughts within Thoughts
within Thoughts within Thoughts to Rethink with me"

"I AM Entertaining Maintaining Sustaining and Supporting Divine
Love Divine Harmony Divine Gratitude Divine Forgiveness Divine
Light Divine Life Divine Happiness Divine Joy Divine Wisdom
Divine Oneness Divine Tranquility Divine Humility Divine Inner
Peace I AM Rethinking Rethink I AM (full birth name) I AM"

Stillness sets in after saying this affirmation Sit with what you just said and believe what you just said

Procrastination

To control your limiting Thoughts is difficult Once the Thoughts are there is no limitation

"I AM owning and rethinking my limiting addiction
Thoughts I AM"

"I invite my history of limiting Thoughts to come into harmony
with disharmony I AM"

"I AM in harmony with my limitations I AM"

"I AM my limitations I AM"

"I AM fulfilling my evolutionary self-correcting Thought Creations
I AM (full birth name) I AM 24/7"

Memory

These affirmations focus on long and short term memory   Secure your resonance before dowsing for accuracy

"I AM every Thought I AM"

"I AM inviting long and short term memory to come into
Harmony with Thoughts throughout my Evolutionary Divine
Thought Creation I AM (full birth name) 24/7 I AM"

"I AM the efficiency and effectiveness of my short and long term
memory I AM 100%"

(check percentage)

If your percentage is not 100 repeat the following and then recheck your percentage   The following can also be said once a day

"I AM inviting Thoughts within Thoughts within Thoughts to
Rethink with me I AM"

"I AM inviting Thoughts within Thoughts within Thoughts
within Thoughts within Thoughts within Thoughts to rethink
with me I AM"

"I AM inviting Thoughts within Thoughts within Thoughts within
Thoughts within Thoughts within Thoughts within Thoughts
within Thoughts within Thoughts to rethink with me I AM"

"I AM (full birth name) I AM"

Lost Item

Often a lost item is not visible because I AM not vibrating at the same frequency as the item   After securing my resonance I invite myself to come into harmony with the lost item   Forget about it and it will show up

Helping the Planet

Joachim has discovered several very powerful affirmations that have important applications   Repeat the following affirmation 3 times 6 times and then 9 times at least once a day to harmonize countries   Yes countries Dowse a specific location to see if it is in harmony   If there is a "No" say this affirmation and dowse again to see the difference Remember to Secure Your Resonance first for dowsing accuracy

"I AM inviting every Thought Creation to please rethink with me

Divine Christ God Collective Consciousness I AM thanking you
for every blessing of Every Thought Creation on this wonderful
planet throughout my Thought Creation I AM (full birth
name) I AM"

The next set of affirmations is particularly important at this time of disharmony on the planet It could be repeated as often as every hour, but saying it morning noon and night would greatly help the planet

"I AM inviting Thoughts within Thoughts within Thoughts to
rethink with me"

"I AM inviting Thoughts within Thoughts within Thoughts within
Thoughts within Thoughts within Thoughts to rethink with me"

"I AM inviting Thoughts within Thoughts within Thoughts within
Thoughts within Thoughts within Thoughts within Thoughts
within Thoughts within Thoughts to rethink with me"

"I AM Entertaining Maintaining Sustaining and Supporting Divine Love Divine Harmony Divine Gratitude Divine Forgiveness Divine Light Divine Life Divine Happiness Divine Joy Divine Wisdom Divine Oneness Divine Tranquility Divine Humility Divine Inner Peace I AM Rethinking Rethink I AM (full birth name) I AM"

After repeating the above affirmation harmonizing the thoughts around me I invite those thoughts to place themselves equally spaced around the planet Earth to harmonize other thoughts and repeat the affirmation again

"I AM inviting Thoughts within Thoughts within Thoughts to Rethink with Me"

"I AM inviting Thoughts within Thoughts within Thoughts within Thoughts within Thoughts within Thoughts to Rethink with me"

"I AM inviting Thoughts within Thoughts within Thoughts within Thoughts within Thoughts within Thoughts within Thoughts within Thoughts within Thoughts to Rethink with me"

"I AM Entertaining Maintaining Sustaining and Supporting Divine Love Divine Harmony Divine Gratitude Divine Forgiveness Divine Light Divine Life Divine Happiness Divine Joy Divine Wisdom Divine Oneness Divine Tranquility Divine Humility Divine Inner Peace I AM Rethinking Rethink I AM (full birth name) I AM"

Health

Property

We live in an environment and are influenced by it   If the property is out of harmony it is difficult to heal   Our bodies need a vibrational frequency that supports it in order to heal   Inviting myself to come to the correct life experience to dowse the property is a start   Joachim advises to go to the correct vibrational frequency to correct what needs to be corrected rather than "clearing" a property

"I AM in harmony with this house (or property)" (Y/N)

(Give the street address if possible)

"This property is in Harmony with Universal Intelligence" (Y/N)

"This house (or property) is in harmony with me" (Y/N)

(To take it to another level, you might check every room to see if it is in Harmony)

If the dowsing tool indicates "No" to the above statements then interfering Thoughts are presently active   State the following to correct

"I Energize Harmonize Neutralize Normalize Polarize
the property"

Consider saying "I AM 100% I AM" before you enter your home to avoid bringing in interfering Thoughts

It is also possible to check the property by viewing it from above or across the street telepathically

Physical Problem

"Giving a label to a dysfunction helps to solidify the dysfunction. The dysfunction then becomes a part of the body, part of you   It will then have its own life" says Jerry Gin

Do not empower the condition by saying "I have (name of illness)" Instead refer to the dysfunction as "an *activity* of (name of illness) in my physical body"

To find the interfering vibration dowse the following

"This illness I experience in my body is self-induced" (Y/N)

If "Yes"

"It is generated by Thoughts of my right brain or left brain" (Y/N)

If "Yes"

"I AM inviting myself to give myself permission to go to the
correct vibrational frequency to correct the situation"

If the illness is not self-induced someone needs help   Dowse to see if it is a friend family member acquaintance or Thought (See Dowsing section for more information in Part 1 of this chapter)

(Experienced dowsers can work the problem to a deeper level finding out what lifetime the Thought was created in while not forgetting to ask if there is something they missed or the client missed)

Below are affirmations that can be dowsed to assist in healing The idea is to find out where the Thought has come from that is manifesting as a health problem and correct it   I invite the client to read and Rethink this with me

If the dowsing tool gets a "No" when dowsing the following then a correction is needed which is stated below

1. "I AM inviting myself to give myself permission to come to the correct vibrational frequency to correct the situation"

2. "I AM owning and rethinking my (name of illness) historic limiting Thoughts I AM"

3. "There is an activity of (name of illness) presently active in my physical body"

   "There is an activity of (name of illness) presently active in my etheric body"

   "There is an activity of (name of illness) presently active in my emotional body"

   "There is an activity of (name of illness) presently active in my mental body"

   "There is an activity of (name of illness) presently active in my spiritual body"

   "There is an activity of (name of illness) presently active in my stealth body"

(This helps you determine where the problem is manifesting)

4. "I AM Rethinking Rethink I AM (full birth name) I AM"

5. "I AM inviting Thoughts of angels I AM"

6. "I AM owning and rethinking my Thoughts I AM"

7. "I AM amazing (full birth name) I AM"

8. "I AM 100% (full birth name) I AM"

9. "I AM inviting Thoughts to Rethink with me I AM"

10. "I AM owning and Rethinking my history of addictive limiting thoughts I AM"

11. "I AM coming into harmony 100% with everything that is I AM"

12. "My Left Brain and my Right Brain join together with my Heart I AM"

13. "I AM inviting my history of limiting Thoughts to come into Harmony with Disharmony I AM"

14. "I AM inviting Thoughts to come into Harmony with Disharmony I AM"

15. "I AM inviting myself to come to the correct vibrational frequency of my First Thought Creation to correct what needs to be corrected I AM"

16. "I AM Inviting my original soul, mind, and spirit to come back into my body I AM"

"I AM the efficiency and effectiveness of my physical life my etheric life my emotional life my mental life my spiritual life my stealth life I AM"

(Measure and check the percentage of each level. Correct if it is not 100% and remeasure until it is 100%)

To correct

"I AM inviting myself to come to the correct vibrational frequency of my first Thought Creation I AM"

"I AM inviting myself to give myself permission to come to the correct vibrational frequency to correct my Thought creation I AM"

"I AM inviting the interfering vibrational frequency to repeat the harmony Affirmation"

"I AM inviting Thoughts to come into harmony with disharmony 100% I AM"

Invite (the client) to come to the correct vibrational frequency of their first Thought Creation to read and rethink with me the following

"I AM in Harmony with my physical body 100%" (Y/N)

"I AM in Harmony with my etheric body 100%"

"I AM in Harmony with my emotional body 100%"

"I AM in Harmony with my mental body 100%"

"I AM in Harmony with my spiritual body 100%"

"I AM in Harmony with my stealth body 100%"

Check percentage   If not 100% say

"I AM in Harmony with my _____ body" (the body that
is not 100%)

(say this 3 times pause 6 times pause then 9 times)

Now dowse to see if the efficiency and effectiveness of the physical etheric emotional mental spiritual and stealth body is 100%

If "No" repeat any of the harmony affirmations (Harmony section) and recheck

Pain or Discomfort

Dowse to see what is causing the pain or discomfort

"This pain my body is experiencing is generated by my
Thoughts" (Y/N)

"This pain my body is experiencing is generated by someone
else's Thoughts" (Y/N)

If "Yes" to either statement, correct by saying the following

"I AM inviting my Thoughts to come home and be with me
100% 24/7"

"I AM self-correcting my disharmony Thoughts I AM"

"I AM the disharmony pain I AM"

Cells

Each cell has consciousness and tries to please   Don't say I have an illness say I was diagnosed with an illness   There can be a misunderstanding of our communication between my Thoughts and my cells   A misinterpretation of language between the cells in my body and myself

"I AM inviting every Thought in every cell to please read and
Rethink with me the Harmony Disharmony Affirmation"

Immune System

Check the efficiency and effectiveness of your immune system

"I AM in Harmony 100% with my immune system" (Y/N)

"My immune system is in Harmony 100% with my cells" (Y/N)

To correct say

"I AM inviting Thoughts within Thoughts within Thoughts to Rethink with Me"

"I AM inviting Thoughts within Thoughts within Thoughts within Thoughts within Thoughts within Thoughts to Rethink with me"

"I AM inviting Thoughts within Thoughts within Thoughts within Thoughts within Thoughts within Thoughts within Thoughts within Thoughts within Thoughts to Rethink with me"

(I AM inviting all the thoughts around me who want to be healed)

"I AM inviting Thoughts to come home and rethink with me 100% I AM"

(Say 3 times then pause 6 times then pause then 9 times)

Check the efficiency and effectiveness of the immune system again
It should be at 100%

Healing Affirmations

If not feeling well I invite my Thoughts to self-correct

"I AM inviting Thoughts to come home and rethink with me 100% I AM"

"I AM in harmony with my own disharmony Thought creation I AM"

Do this as often as possible

Yawning

Yawning means there is too much disharmony in the body, and you are being drained

Secure Your Resonance (*see* Part 1 of this chapter) and invite myself to come home and be with me 100%.

Depression

> "I AM in Harmony with my own Disharmony 100% I AM"

Stress

Stress is not the event it is your reaction to the event

> "I AM in Harmony with Stress I AM"

Age

An invitation to slow down aging

> "I AM aged enough I AM"

Death

To assist someone in transitioning requires making sure they are "home"

Dowse to see if they are in their body 100%   If not invite them to come "Home" and be with themselves

Dowse if they are ready to graduate from this world life experience Dowse what percent is ready   Dowse if they have anything left to do in this lifetime

Suicide A person is not in their body and the Karmic ramifications go to the Thought Creation

Weight

Check assimilation and absorption of nutrients

Check on physical body digestion and elimination

If assimilation is not 100% it can cause weight gain

Who am I eating for

Dowse to see which part of self wants more (Physical Etheric Emotional Mental Spiritual Stealth)

To correct

> "I AM in harmony with my _____ body I AM"

(Fill in the blank with the level which was not in harmony Physical Etheric Emotional Mental Spiritual Stealth)

> "I AM addicted to food I AM"

"I AM donating the extra pounds to someone who needs
them I AM"

"I AM Zero appetite I AM"

"I Am Zero Limitations I AM"

"I AM owning my own Thoughts I AM"

"I AM Evolving I AM"

(Put the affirmation on your refrigerator)

Before eating or drinking put the plate or cup between the hands for 45 seconds to harmonize it with the person's energy field

When eating at a restaurant say

"I AM inviting the growers and servers to come into harmony
with their own disharmony"

Dowsing

Questions vs Statements

Traditional mental dowsing asks questions   Once the initial questions are asked Joachim Wippich has a unique approach   He transforms the question into a statement   "Asking questions gives you questionable answers" he shares with his students   By making a statement instead the statement is either correct or incorrect   This approach provides for more accurate dowsing

He also teaches that when dowsing the pendulum gives you a correct or incorrect motion   Referring to this motion as true or false is subjective because what is true for you may not be true for me   State I will use "Yes"/"No" as an indicator for the accurate / inaccurate not true / false when programming a pendulum

The first step when dowsing should be to **Secure Your Resonance** The 3 steps of the process ensure dowsing accuracy when your "I AM" is 100%   This means there is no room for interfering Thoughts to come

The affirmations to Secure Your Resonance have been described in Part 1 of this chapter   Please refer to that part of the chapter

Relationships

To assist in securing a relationship I invite the person to help me fulfill my evolutionary destiny   Make sure he/she is in their physical body by dowsing to see if they are "home" 100%   Invite them to come "home" and stay with their physical body 24/7   Remember when I invite I allow for free will

"I AM inviting my Thoughts to Rethink with me especially for my correct Partner I AM"

"I AM ready to meet the correct partner I AM"

Relationship Situation

I Secure my Resonance first then bring myself to the correct vibrational frequency to deal with the situation

"I AM inviting myself to come to the correct vibrational frequency to deal with the situation"

"The relationship between myself and (he/she) is perfect" (Dowse if Y/N)

"The relationship between (he/she) and myself is perfect" (Dowse Y/N)

"I AM inviting (he/she) to come home and stay in their body 24/7" (Dowse Y/N)

"There is tension between myself and (he/she)" (Dowse Y/N)

"There is tension between (he/she) and myself" (Dowse Y/N)

"The is a disharmony Thought causing tension between us" (Dowse Y/N)

I AM Inviting interfering Thoughts creation to come home and be with me on every level I AM

"I AM in harmony with my physical body I AM" (Y/N)

"I AM in harmony with my etheric body I AM" (Y/N)

"I AM in harmony with my emotional body I AM" (Y/N)

"I AM in harmony with my mental body I AM" (Y/N)

"I AM in harmony with my spiritual body I AM" (Y/N)

"I AM in harmony with my stealth body I AM" (Y/N)

To correct

"I AM Inviting interfering Thoughts creation to come home and be with me on every level 100% of my I AM being"

"The relationship between he/she and (client) is in harmony with each other"

"The relationship between (client) and he/she is in harmony with each other"

"I AM Inviting them to come to the correct vibrational frequency where they can communicate with each other"

"I AM inviting he/she to come into harmony with (client)"

"I AM inviting (client) to come into harmony with he/she"

Family

I secure my resonance (my I AM is 100%) then I invite them to rethink with me the harmony affirmations (Harmony section)

Helping Others

After Securing Your Resonance dowse:

Who: "I AM...," followed by anyone else involved (who you are dowsing for)

What: Invite to be in 100% Divine harmony with disharmony

When: Usually now, but you may go beyond this timing

Where: This Universe, this Earth, this home, this place

How: By bringing the person into 100% harmony

These steps can be a useful guide when helping someone Check each step with your dowsing tool for a "Yes" motion Correct if "No"

1. "I AM inviting myself to come to the correct vibrational frequency of my first Thought Creation."

2. "I AM in harmony with my physical body I AM"

   "I AM in harmony with my etheric body I AM"

   "I AM in harmony with my emotional body I AM"

   "I AM in harmony with my mental body I AM"

"I AM in harmony with my spiritual body I AM"

"I AM in harmony with my stealth body I AM"

3. Invite Divine harmony spirits (angels, guides) to assist in dowsing

4. When helping someone dowse about interfering Thoughts The three statements below indicate the degree of urgency of someone seeking help with "seeks" being less urgent than "requests" and "needs" being of greatest urgency

"Someone seeks my help"

"Someone requests my help"

"Someone needs my help"

If "Yes," check if it is a relative, friend, ancestor, or entity (Thought)

5. "I AM inviting myself to come to the correct vibrational frequency to correct what needs to be corrected"

6. "I AM inviting the disharmony Thought to come to the correct vibrational frequency to communicate with me"

7. "I AM Inviting the disharmony Thought to accept the gift of the harmony affirmation"

8. "I AM Rethinking Rethink I AM"

"I AM in harmony with my knowledge I AM"

9. "I AM Inviting (the client) to invite the interfering vibrational frequency to repeat the harmony affirmation"

10. "There is something I missed" (Y/N)

"There is something (the client) missed" (Y/N)

Repeat "I AM Rethinking Rethink I AM" until you get "No"
To check your accuracy after dowsing, state the following

"I AM Divine I AM"

"I AM my Divine Thoughts throughout my evolutionary Thought creation I AM"

"I AM (full birth name) I AM"

If you get "No" to any of these statements the dowsing is incorrect

Pets

Invite pet Thoughts to repeat the harmony affirmations   Everything has consciousness and invitations can be telepathic

> "I AM inviting Thoughts to come home and rethink with me 100% I AM"

Travel

This affirmation is asking your guides to make your trip safe and easy

> "I AM inviting 100% Divine spiritual guides to help me make an easy and uneventful trip"

Words

Every word has a vibrational frequency   Understanding that there are limiting words is very important   I can put my hand on a dictionary and dowse if I AM evolving by reading it   I will confirm that there are limiting words in it that have a negative vibrational frequency by getting a "No" response to the evolving statement

Try dowsing a statement with the word "All" and then test "Every" as a replacement   The dowsing tool should give you an incorrect motion with "All" and a correct motion with the word "Every"   This is an example of how to test for a limiting frequency

"All" is a broadly encompassing word which may not have complete truth whereas "Every" can individualized to select segments

Use of the word "So" has increased in our society and Joachim finds that will knock someone out of harmony when they use it   Dowse "I love you so much" or "Thank you so much"   They both have the word "so" which indicates a limitation   Replace "so" with the word "very" and dowse again to see if it is an evolving statement

> "I AM using limiting Thoughts in my conversations or vocabulary" (Y/N)

If "Yes," dowse what percentage and correct until you reach a low of at least 5 or 6%

To correct

"I AM Rethinking Rethink I AM"

Repeat 3 times pause 6 times pause and then 9 times

Blessings

Instead of saying "God bless you" rephrase to "May God bless you" which allows free will

Thank you for allowing me to share these wonderful teachings of Joachim Wippich with you. May they bring you peace and harmony.

APPENDICES

Working with a Pendulum:
Making a Neutral Pendulum / Finding Your Personal Wavelength

How to Make a Neutral Pendulum
(as shown in Chapter 3)

Recipe for making your own neutral pendulum from flour and salt:
The dough:

1. Mix ¼ cup salt and ½ cup flour with a little less than ¼ cup of water. (This is a consistency that works for me.)

2. Knead the dough.

3. Take enough dough to form a 1-inch ball (or any other shape you want).

4. Cut a string length of about 7 inches.

5. Tie a knot on the end of the string and insert it into the dough. (The knot helps prevent the string from pulling out of the pendulum after you form it.)

6. Then form the pendulum, keeping the string in the center of the ball of dough so that the pendulum is balanced. (You can also place the pendulum on some flour as a cushion to prevent it from deforming while drying.)

The string:

For string (typically sold as twine), #12 size is a good thickness. Don't get tar-coated string; it is too stiff. Nylon string is also problematic: it tends to unwind, so you have to coat the ends with glue. Most personal wavelengths run around 1.5 inches (that is, the length of the string that corresponds to your resonant energy), so for practical purposes you don't need an excessive string length.

Let the pendulum dry:

1. Let the pendulum dry on its own for a few days.

2. At that point, you can (if you want to) tie a loop at the end of the string and place the loop on your pinky finger so that the excess string doesn't get in the way. But since you don't need an excessive length (as mentioned above), this is optional.

3. If you want to eliminate the rougher dough feel of the pendulum, you can give it a light coat of polyurethane (optional).

Below are a few pendulums I made using the above procedure.

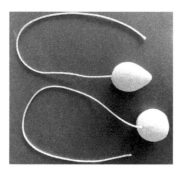

FIGURE 12. Pendulums made of salt, flour, water + string.

APPENDIX II

Exercises

From Chapter 2: I AM, The Basis of a New Spirituality

Blessing Water

1. Taste some tap water in a glass. Then put your hands around the glass of water and bless it with love and gratitude for 1 minute.

2. Then slowly sip some of the blessed water.

3. How does it taste? Is it better than the original water? Does it taste more like good-quality spring water?

From Chapter 3: Resonance, Subtle Energies, and Tools for Their Detection

Using a Vial as a Pendulum

With radiesthesia, we merely watch the pendulum once a to-and-fro motion has been initiated, and see if there is a clockwise rotation. If there is, it indicates the presence of resonance with the subtle energy. You can use a pendulum to see if there is resonance with almost anything (e.g., water, minerals, or yourself).

To illustrate this, I made a pendulum out of a plastic vial with a cap, and attached a string to the cap (see Figure below). If you have a plastic vial (or a container on which you can attach a string) as shown in the figure below, you can do the experiment too.

Example 1:

1. Fill the vial, converted into a pendulum, with water. The pendulum is now constructed to be in resonance with water.

2. When the pendulum is placed over a glass or vial containing water, it will go into resonance with the water in the glass or vial.

3. After you have initiated a to-fro-motion of the pendulum, the pendulum will rotate clockwise because of resonance with the glass or vial of water.

Exercise 2:

1. Place an aspirin in a vial converted into a pendulum.

2. When the pendulum is placed over aspirins on a table, it will go into resonance with the aspirin.

3. After you have initiated a to-and-fro motion of the pendulum, the pendulum will rotate clockwise because of resonance with the aspirin on the table.

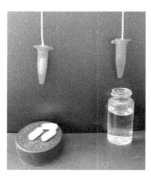

FIGURE 9. Pendulums made from vials with an aspirin and with water. These pendulums will rotate clockwise when they are held above either water or aspirin pills, demonstrating your ability to detect resonance using a focusing device (the pendulum).

Practicing for Resonance with an Easy-to-Make Metal Pendulum

Since you may not have a vial in which you can attach a string to carry out the above examples with water or aspirin, you can carry out another exercise.

1. Tie a string to a metal nut. This creates a pendulum that would be able to come into resonance with another metal nut, washer, or bolt of the same type of metal.

2. Hold the string about 2 inches from the nut.

3. Practice a while to let your muscle memory get familiar with a to-and-fro motion, clockwise rotation, and counterclockwise rotation. Then create a to-and-fro motion with the pendulum over a nut, washer, or bolt made from the same metal as the nut on the string. As resonance is achieved, the pendulum will rotate in a clockwise direction.

4. If you now move the pendulum over another type of surface (e.g., wood table) and again start the to-and-fro movement, the pendulum will have just the to-and-fro motion, indicating that no resonance was achieved.

Below is a picture of a pendulum made from a nut and a washer, which can be placed on a table to check for clockwise rotation. If you have problems doing this exercise, the section after learning how to make a neutral pendulum ["Preparation Exercises Before Working with a Pendulum"] describes how to overcome common blocks in this type of exercise and when using pendulums.

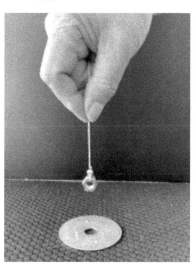

FIGURE 10. A pendulum made with a string and a nut. The nut is metal and the washer below the nut is metal. Because of resonance between the two similar metals, the pendulum swings clockwise once the to-and-fro motion has been initiated to overcome inertia.

Preparation Exercises Before Working with a Pendulum

Feeling Energy Sensations

1. Stare at your fingertips and mentally ask to feel the sensations in your fingertips. Do this for about a minute and experience whatever sensations you feel. If you are right-handed, be sure to also try this exercise with the right hand (although I find that my left hand is more sensitive than my right hand).

2. Then place your fingers about 7 inches above a surface (table, chair) and see what sensations you feel in your fingertips. This gets you used to feeling energies.

3. Afterwards, you can put your left hand on your right shoulder and brush down with the left hand to the right hand. This reminds the body that you will be moving energy down the right hand to the right fingertips (for a right-handed person).

Making Sure Your Polarity Is Not Switched

Occasionally a person's energy polarity is switched, so pendulums that should rotate clockwise will rotate counterclockwise. A simple physical movement usually corrects this phenomenon.

1. While standing, lift up your left knee high enough to tap the knee with your right hand.

2. Then lift up your right knee and tap with your left hand.

3. Repeat this motion 5 or 6 times.

It is like marching in place but touching the opposite knee each time, and at a speed similar to marching/walking. This usually corrects polarity switches.

Training the Muscle Memory of Your Fingers with a Pendulum

Most people do not know the feel of a pendulum when it is rotating clockwise, counterclockwise, and making a to-and-fro motion.

1. Using the index finger and thumb, pinch the string of the pendulum about 2 inches from the dough ball.

2. Play with the pendulum and make it rotate clockwise and counterclockwise, let it move to-and-fro, away from you and towards you, and also move left and right.

3. Practice these series of motions with the pendulum until you are comfortable that your body remembers them.

Getting the Feel of a Pendulum Pulling in a Clockwise and Counterclockwise Manner by Testing Battery Polarity

1. Take out any battery (e.g., AA, C) and place it on its side on a table.

2. With your index finger and thumb, hold the string of the pendulum about 2 inches from the pendulum weight (the dough).

3. Start a forward-and-backward to-and-fro movement over the positive end of the battery.

4. Once the motion has been initiated, let the pendulum swing on its own (no attempt by you to make a to-and-fro motion). Just allow the pendulum to move freely on its own. You will notice that the pendulum, on its own, will rotate clockwise.

5. Bring the pendulum to the middle of the battery and start the to-and-fro motion. You will notice that it will just continue its to-and-fro motion.

6. Now bring the pendulum to the negative end of the battery and start the to-and-fro motion. If you allow the pendulum to move freely on its own, the pendulum will rotate counterclockwise on its own.

This exercise allows you to "feel" how a pendulum will move on its own without conscious control on your part. Your body "knows" the polarity of the battery by the gravitational pull of the positive pole making the pendulum rotate clockwise, or the magnetic pull of the negative pole making the pendulum rotate counterclockwise. That subconscious information is transmitted through your arm to your fingertips, and the rotation of the pendulum occurs naturally without conscious thought on your part.

Below is a picture of a battery standing up (rather than on its side, as described above). When you start the forward-and-backward motion of the pendulum, the pendulum will rotate clockwise over the positive pole and counterclockwise over the negative pole. Dr. Robert Gilbert at the Vesica Institute has an excellent instructional YouTube video reviewing how to use a neutral pendulum: *https://youtu.be/rklULOn8p0w*.

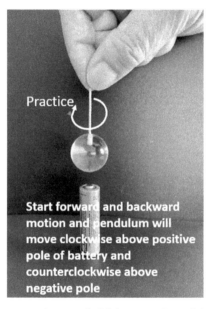

FIGURE 13. For practice, when you hold the neutral pendulum above a battery and initiate the to-and-fro movement, the pendulum rotates clockwise over the positive pole and counterclockwise over the negative pole.

Detecting Resonance of Colors

(See Figure 15 below for detecting the wavelength of red):

1. Assemble an object that has a color (e.g., colored paper) on a flat surface. It could be red, blue, or any other color.

2. Take your neutral pendulum, and hold the string between your index finger and thumb, with the weight next to the thumb/

index finger (i.e., put your thumb and index finger all the way down the string, just above the weight).

3. Start the forward-and-backward to-and-fro motion of the pendulum as you gradually lower the weight over the colored object.

4. Once the to-and-fro motion of the pendulum has been initiated, you no longer are consciously trying to make a to-and-fro motion. You are allowing the pendulum to move on its own.

5. The pendulum will change to a clockwise rotation at the wavelength of resonance with the color. That wavelength (string length) will now rotate clockwise over that color on any object. The length of string in which resonance occurs is approximately 1.5 inches. It usually takes me about 5 to 10 seconds to lower 1.5 inches on the pendulum string.

   Every different color will have its own wavelength, but they all are close to the 1.5 inches and are easily determined to be different from one another, since the string length that causes the pendulum to move clockwise for red is different from the string length that causes rotation for any other color. If you had tested for the string length for the color red, all other objects in your house that are red will also cause the pendulum to rotate clockwise. For any other color, the pendulum will just have the to-and-fro motion.

6. You can prove this for yourself by now bringing the neutral pendulum over another object with that same color. It will rotate clockwise. Then hold the pendulum over an object with a *different* color. The pendulum will just have the to-and-fro motion and will not move clockwise.

   Feel free to repeat this experiment with other colors until you are comfortable with carrying out this exercise and you understand the concept of resonance with colors.

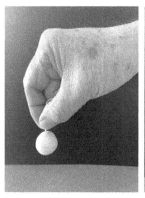

FIGURE 14. Wavelength of the color red. Using a neutral pendulum, gradually lower the pendulum by sliding up the string with the fingers over the red surface. Overcome inertia by creating a to-and-fro movement. When the string length reaches the wavelength of red, the pendulum will rotate clockwise. The pendulum will not rotate clockwise over other colors but will do so over any other object that is red.

Practical Uses of Personal Wavelength

Foods

1. Go to the refrigerator and pick out a sampling of foods.

2. Slowly release the string of your neutral pendulum over the back of your hand and initiate a to-and-fro movement (see instructions above on finding your personal wavelength).

3. Once the pendulum is moving to and fro, let it move on its own.

4. At around 5-10 seconds, the pendulum weight will have dropped around 1.5 inches. Usually, within the 5–10 second and the 1.25–2.0-inch range, the pendulum will start rotating clockwise on its own. The pendulum typically will also rotate a little beyond when the pendulum starts rotating, since there is a small range that is the resonant string length. Beyond that length, the pendulum stops rotating. The range in which your pendulum rotates is your personal wavelength. (The speed of rotation is also an indication of your vibrational level.) At twice the string length for personal wavelength, the pendulum will again rotate, since that is the next resonant octave. However,

it is much easier to work with the first octave, since the string length is shorter.

5. Keep your thumb and index finger at the personal wavelength distance.

6. Put your non-dowsing/non-radiesthesia hand over a food. (Your left hand, if you are dowsing right-handed; your right hand, if you are dowsing with your left hand.) By doing so, through your intention, you have linked yourself to the food in question. Although you could be touching the food, you do not have to. Just having your hand over the food connects you to the food.

7. Start the to-and-fro motion of the pendulum and then observe its motion.

   - The pendulum will pull into a clockwise rotation if the food is in resonance with you (i.e., good for you).
   - If the pendulum remains in a to-and-fro motion, the food is neither good nor bad for you. It is not benefiting you but it is also not harmful.
   - If the pendulum rotates counterclockwise, you should stay away from that food since it is detrimental for your body (i.e., not in resonance with you).

8. Repeat this procedure with other foods that you have brought out of the refrigerator. Typically, foods that have been grown in the presence of glyphosate (a detrimental herbicide sometimes used in non-organic farming) will cause the pendulum to rotate counterclockwise. If a food is contaminated, the pendulum will rotate counterclockwise. Do not eat those foods. If you have some older food in the refrigerator, test them to see if the pendulum goes clockwise or counterclockwise. This indicates whether you can still eat the food or should discard it (i.e., if the food is spoiled).

9. If you have any type of artificial sweetener, test it with the personal wavelength procedure, above. The pendulum will likely turn counterclockwise.

10. Use the above procedure whenever you go shopping. Perhaps counter to your expectations, non-organic foods are often okay for you (the pendulum will rotate clockwise) and some organic foods may sometimes cause the pendulum to go counterclockwise (i.e., not be good for you). What's behind this? You do not necessarily know the history of the food. It could have been grown in fields that previously used a detrimental chemical, or there could be contamination in the water supply. Testing is the best way to determine if a food is good/healthy for you or not healthy for you.

Supplements and Drugs

1. Take out the drugs and supplements you normally use, and put them on a surface (e.g., a table).

2. Slowly release the string of your neutral pendulum over the back of your hand, and initiate a to-and-fro movement (see instructions above on finding your personal wavelength).

3. Once the pendulum is moving to-and-fro, let it move on its own.

4. After around 5-10 seconds, the pendulum weight will have dropped around 1.5 inches. Usually, within the 5–10 second and the 1.25–2.0-inch range, the pendulum will start rotating clockwise on its own. Beyond that length, the pendulum will stop rotating. The range in which your pendulum rotates is your personal wavelength. (The speed of rotation is also an indication of your vibrational level.)

5. Keep your thumb and index finger at the personal wavelength distance.

6. Put your non-dowsing/non-radiesthesia hand over a drug or supplement. By doing so, through your intention, you have linked yourself to the drug or supplement in question. You could be touching the drug or supplement, but you do not have to. Just having your hand over the container holding the drug

or supplement connects you to the drug or supplement. You can prove this to yourself by testing the drug or supplement in the container, and by taking the drug out of the container and putting your hands on the drug or supplement.

7. Start the to-and-fro motion of the pendulum over your hand (which is also over the drug or supplement), and then observe the motion of the pendulum.

   - The pendulum will pull into a clockwise rotation if the drug or supplement is in resonance with you (i.e., good for you).
   - If the pendulum remains in a to-and-fro motion, the drug or supplement is neither good nor bad for you. It is not benefiting you, but it is also not harmful.
   - If the pendulum rotates counterclockwise, the drug or supplement is detrimental for your body (i.e., not in resonance with you).

   You can also obtain an idea of dosage for drugs and supplements. You might vary the number of pills being tested (e.g., 1, 2, or 3 pills) and see what happens with pendulum rotation as you increase the number of pills. Be aware that some drugs may not be good for you but your doctor wants a certain outcome in spite of the problem the drug may cause. In this case, you may want to talk to your doctor about side effects of the drug and then determine what you might want to do. It's somewhat similar to a dessert that may be very high in sugar, to which the pendulum will rotate counterclockwise. You may still decide to eat the dessert, knowing that you are not eating an excess of dessert every day.

8. Repeat this procedure with other drugs and supplements that you have brought out. You will now know which drugs and supplements are in resonance with you (beneficial).

EMF Radiation (Electromagnetic Field)

1. Determine the string length of your personal wavelength. (See directions earlier in this chapter.)

2. Once you are at your personal wavelength, initiate the to-and-fro motion of the pendulum and connect to a source of EMF. At a minimum, you should test your cell phone, cordless phone, router for your WiFi, and computer using WiFi. You connect to the phone, router, or computer by just putting your hands over the electronic device and seeing what happens to the rotation of the pendulum.

3. You will then see that EMF will cause a counterclockwise rotation. There are BioGeometry tools (strips) for the electronic devices that will harmonize the EMF.

Stones (Minerals, Gems), Essential Oils, and Homeopathic Remedies

Minerals, gems, essential oils, and homeopathic remedies are often sold for their medicinal value. However, there is always the question of whether or not the particular stone or oil is truly beneficial *for the particular ailment you are trying to treat*. The best way to determine this is by testing, and the best test is with the personal wavelength method.

Follow the instructions given in the Personal Wavelength Exercise and see if the particular remedy will cause the pendulum to rotate clockwise for you (i.e., is in resonance with you).

You should also be aware that many minerals are irradiated to enhance the color of the crystals. From the perspective of giving beneficial energies, this is very bad and often gives off negative energies, causing the pendulum to rotate counterclockwise, making it unhealthy for you. Testing before you buy is essential.

From Chapter 4: My Experiences in Feeling Qi/Subtle Energies

Feeling Qi

The ability to feel qi is something that develops as you practice feeling the energies. Here is an exercise for feeling qi:

1. Place your hands in front of you, with the palms facing each other, as if you are holding a basketball (about one foot apart).

2. Now move that invisible ball around (back and forth, up and down, in circular patterns) in a pretend Tai Chi exercise. Pay attention to the sensations you can feel between your hands. You might be surprised to feel a tingling sensation or the build-up of warmth between your palms.

3. Let the qi build as you play with that invisible ball of qi for about 5 minutes (i.e., have your palms facing each other about a foot apart).

Seeing Lines or Rays between Your Hands in Front of a Light-colored Wall

Here is an exercise to see if you can see the lines or rays between your hands:

1. Find a blank, light-colored wall (e.g., ivory or light beige, with no designs).

2. Hold your hands up with the palms facing each other about 8 inches apart, as if you are holding an invisible 8-inch ball.

3. Bring your hands to eye level, and look through the empty space between your palms.

4. Move both hands up and down simultaneously; or you can move one hand up a few inches, and keep the other hand steady. The movement allows you to tell the difference between seeing lines and no lines more clearly.

Seeing Beams of Energy from Your Hands in the Dark

You can also try an experiment of seeing beams of energy that emit from your palms or fingers in a darkened room (no lights; just moonlight in the room).

1. Adjust your eyes to the dark, first.

2. Pretend that your palm is a flashlight. See if you can see a line of energy that is emitted by the palm.

From Chapter 5: Subtle Energies—A Developing Story

Entropy and Negentropy Pendulum Spins

Testing the spin of the pendulum with entropy, using evaporating acetone (typically available from hardware stores; if you don't have acetone, you can use fingernail-polish remover).

1. Using a neutral pendulum, hold the string at about 1.5–2 inches from the weight on the pendulum.

2. Over the closed bottle of acetone, start a to-and-fro motion of a neutral pendulum. The pendulum will remain in a to-and-fro motion.

3. Now open the bottle of acetone and pour a little of the acetone into an open vessel (e.g., a cup, bowl, or the cap of the acetone bottle).

4. Start the to-and-fro motion of the pendulum.

5. The evaporating acetone will cause the pendulum to rotate counterclockwise (decay, entropy).

Testing the spin of the pendulum with negentropy, using stretched rubber band:

1. Using a neutral pendulum, hold the string at about 1.5–2 inches from the weight on the pendulum.

2. Place a rubber band, in its relaxed position, on a surface.

3. Hold the pendulum over the rubber band.

4. Start a to-and-fro motion. The pendulum will remain in the to-and-fro motion.

5. Put one end of the rubber band on any stationary protrusion (nail, screw, etc.) and stretch it out.

6. Start the to-and-fro motion of the pendulum over the stretched rubber band.

7. The stretched rubber band will cause the pendulum to rotate clockwise (negentropy).

From Chapter 6: The Source of Spirals

The Golden Ratio

Golden Ratio (φ): Phi=1.618

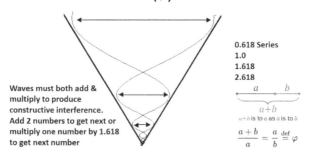

0.618 Series
1.0
1.618
2.618

Waves must both add & multiply to produce constructive interference. Add 2 numbers to get next or multiply one number by 1.618 to get next number

$a+b$ is to a as a is to b

$$\frac{a+b}{a} = \frac{a}{b} \overset{\text{def}}{=} \varphi$$

FIGURE 30. Dan Winter has found that compression waves that follow the golden ratio cause constructive interference, which results in other waves following a similar pattern (***phase conjugate***) to compress in the same manner. He defines the compression pattern as ***implosion***. Spirals are activated by tracing them with a pencil. When activated, BG3 and 3-6-9 is found to be associated with the spiral curve. You can prove this by following the spiral with a BG16 or a 3-6-9 pendulum (the tools to detect BG3 and 3-6-9).

You can prove this by following the spiral with a BG16 or a 3-6-9 pendulum (the tools to detect BG3 and 3-6-9).

From Chapter 10: Who We Are—Moving and Copying Subtle Energies/Fields

Measuring Other Subtle Energy Qualities

The experiment [with moving energies by sheer intention] then expanded to include other subtle energy qualities that we could measure. So I looked at the subtle energy of various items.

In one example, I determined the energy of an essential oil—in this case, lavender (it has horizontal infrared energy)—and moved that energy across the room.

Another experiment was carried out to determine the wavelength of a color using a neutral pendulum, and then moving that subtle energy color quality to another location. The color quality had indeed moved to the new location. This is an easy experiment, which you can try.

Testing for detection of subtle energies on a photocopy of various surfaces where intention was used to move energy onto different surfaces (wood, CD, stone, paper)

A question that I decided to test was to see if the subtle energies on the various surfaces could be photocopied. I put the wood, CD, stone, and paper onto the surface of a photocopying machine. The photocopied image of the materials also had the same level of subtle energy as the original 3-6-9 aum array of array. You can test the BG3 and 3-6-9 of the photograph and see if you can detect BG3 and 3-6-9.

Wood, Stone, CD, Paper for Water Activation

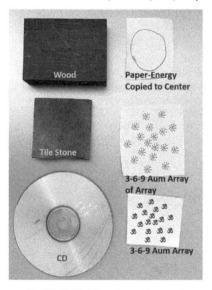

FIGURE 60. The above figure shows the materials (wood, stone, CD, and paper) imprinted with the subtle energies of the 3-6-9 aum array of array, as shown. The 3-6-9 aum array (3 aum symbols surrounded by 6 aums surrounded by 9 aums) was miniaturized and used to create another 3-6-9 aum array of array. The aum array has 6,400 BG3 units, and the aum array of array has 17,500 BG3 units. A vial of water was placed on each surface and monitored for the kinetics of water activation by following its BG3 and 3-6-9 levels. A vial of water not placed on an imprinted surface was used as a control.

References

Cover illustration

Orgonite pendant designed by John Thompson, Mother Earth Orgone (*http://www.mother-earth-orgone.com/*). Photo by Jerry Gin. The cover picture shows orgonite, using metals and crystals to produce a beautiful pendant with BG3 and 3-6-9 subtle energies of 5000 units on the BG3 Ruler.

Introduction

Williams, Bernard O., Ph.D., "Exploring Multiple Meanings of Subtle Energy," *Subtle Energies & Energy Medicine,* Vol. 20, No. 2 (*https://journals.sfu.ca › seemj › article › download*)

Chapter 1: Applying Scientific Hypotheses to Subtle Energy, Consciousness, and Intention

Backster, Cleve. *Primary Perception: Biocommunication with Plants, Living Foods, and Human Cells.* Anza, CA: White Rose Millennium Press, 2003.

Emoto, Masaru. *The Hidden Messages in Water.* New York: Atria Books, 2005.

Houck, Jack. *https://www.youtube.com/watch?v=wFsfaCIE34s* (on spoon bending).

Kozyrev, Nikolai. His work is described in Claude Swanson's *Life Force: The Scientific Basis,* Volume II of *The Synchronized Universe*, 2nd ed. Tucson, AZ: Poseidia Press, 2011.

Persaud-Sharma, Dharam, and O'Leary, James P. "Fibonacci Series, Golden Proportions, and the Human Biology." *Austin J. Surg.* 2015;2(5): 1066.

Radin, Dean. *Supernormal: Science, Yoga, and the Evidence for Extraordinary Psychic Abilities.* Published by Deepak Chopra, 2013.

Reich, Wilhelm. His work is described in Claude Swanson's *Life Force: The Scientific Basis,* Volume II of *The Synchronized Universe*, 2nd ed. Tucson, AZ: Poseidia Press, 2011.

Swanson, Claude. *Life Force: The Scientific Basis,* Volume II of *The Synchronized Universe*, 2nd ed. Tucson, AZ: Poseidia Press, 2011.

Targ, Russell. "Third Eye Spies" (documentary film).

Volkamer, Klaus. *Discovery of Subtle Matter: A Short Introduction.* North Rhine-Westphalia, Germany: Brosowski Publishing, 2017.

von Reichenbach, Baron Dr. Carl (Karl) Ludwig. His work is described in Claude Swanson's *Life Force: The Scientific Basis,* Volume II of *The Synchronized Universe*, 2nd ed. Tucson, AZ: Poseidia Press, 2011.

Chapter 2: I AM, The Basis of a New Spirituality

Wilcock, David. *The Source Field Investigations: The Hidden Science and Lost Civilizations Behind the 2012 Prophecies*. New York: Dutton, 2011.

Chapter 4: My Experiences in Feeling Qi/Subtle Energy

Bengston, William. *The Energy Cure: Unraveling the Mystery of Hands-On Healing*. Louisville, CO: 2010.

Cory, Caroline. *Superhuman: The Invisible Made Visible* (film). *https://www.superhumanfilm.com/*

Gordon, Richard. *Quantum Touch: The Power to Heal*, 3rd ed. Berkeley, CA: North Atlantic Books, 2006.

Master Choa Kok Sui, *Pranic Healing*. Newburyport, MA: Red Wheel/Weiser, 1990.

Pearl, Eric. *The Reconnection: Heal Others, Heal Yourself*. Carlsbad, CA: Hay House, 2003.

Chapter 5: Subtle Energies—A Developing Story

Bengston, William. *The Energy Cure: Unraveling the Mystery of Hands-On Healing*. Louisville, CO: 2010.

Strashun, Olga. *Subtle Energy: Information to Enhance, Guide, and Heal*. Canada: Strashun Institute.

Swanson, Claude. *Life Force: The Scientific Basis*, Volume II of *The Synchronized Universe*, 2nd ed. Tucson, AZ: Poseidia Press, 2011.

Volkamer, Klaus. *Discovery of Subtle Matter: A Short Introduction*. North Rhine-Westphalia, Germany: Brosowski Publishing, 2017.

PART III: Subtle Energies

Volkamer, Klaus. *Discovery of Subtle Matter: A Short Introduction*. North Rhine-Westphalia, Germany: Brosowski Publishing, 2017.

Chapter 6: The Source of Spirals

Binder, Timothy A. (about Walter Russell's work). *In the Wave Lies the Secret of Creation*. Waynesboro, VA: University of Science & Philosophy, 1995.

Clark, Glenn. *The Man Who Tapped the Secrets of the Universe* (biography of Walter Russell). Eastford, CT: Martino Fine Books, 2011.

Lao and Walter Russell. *Universal Law, Natural Science & Living Philosophy: A Home Study Course*. Waynesboro, VA: The University of Science and Philosophy.

Manek, Nisha J., MD. *Bridging Science and Spirit: The Genius of William A. Tiller's Physics and the Promise of Information Medicine.* Pacific Grove, CA: Conscious Creation, 2019.

Russell, Walter. *The Message of the Divine Illiad*, Vols. 1 and 2. Waynesboro, VA: University of Science & Philosophy, 1971.

_____. *The Secret of Light,* Illustrated edition. Bridger House Publishers, 2018.

_____, *The Universal One.* Waynesboro, VA: University of Science & Philosophy, 1974.

Tiller, William; Dibble, Jr., Walter; and Kohane, Michael. *Conscious Acts of Creation: The Emergence of a New Physics.* Walnut Creek, CA: Pavior Publishing, 2001.

Winter, Dan. *Fractal Conjugate Space & Time: Cause of Negentropy, Gravity and Perception: Conjuring Life: "The Fractal Shape of TIME" Geometric Origins of Biologic Negentropy. http://www.fractalfield.com/conjugategravity/*

Chapter 8: Measuring 3-6-9 Levels of Creation, Torus Symbols, and Creating Powerful Arrays

BG3 ruler. *See: https://biogeometryeurope.com/product/bg3-scale-with-sphere-attachment/* (See image for "BG3 Scale with Sphere")

BioGeometry Foundation Courses: *https://www.biogeometry.ca/courses-events*

Melchizedek, Drunvalo. *The Ancient Secret of the Flower of Life*, Volume 1. Flagstaff, AZ: Light Technology Publishing, 1999.

Persaud-Sharma, Dharam, and O'Leary, James P. "Fibonacci Series, Golden Proportions, and the Human Biology." *Austin J. Surg.* 2015;2(5): 1066.

Chapter 9: The 3-6-9 Energies of GANS, Ormus, and Their Fields

George Wiseman, Brown's gas kits and books: *www.eagle-research.com*

References for GANS:
Graphics and terminology from Mehran Keshe's website: *https://keshe.foundation/about/mehran-t-keshe*

Keshe Foundation Space Ship Institute: *https://keshe.foundation/education*

Explanation by Keshe of his beliefs: *https://www.youtube.com/playlist?list=PLpCKWzA-bp9unXm9drxdDwX82l6QtYqc-*

On the various states of matter and their energy density: *http://www.plasmaproduction.org/ https://magravsplasmaproducts.com/*

References for Ormus:
Barry Carter, on Ormus: *http://www.subtleenergies.com*

Keshe Foundation, on making Ormus: *https://www.youtube.com/ playlist?list=PLpCKWzA-bp9unXm9drxdDwX82l6QtYqc-*

https://keshe.foundation/education

References for Brown's Gas:

George Wiseman's kits and books: *www.eagle-research.com*

Chapter 10: Who We Are—Moving Fields and Copying Subtle Energies/ Fields

Emoto, Masaru. *The Hidden Messages in Water*. New York: Atria Books, 2005.

Manek, Nisha J., MD. *Bridging Science and Spirit: The Genius of William A. Tiller's Physics and the Promise of Information Medicine*. Pacific Grove, CA: Conscious Creation, 2019.

Nazorov, Igor, and Kronn, Yury. Lecture, "Twenty Years of Vital Force Technology: Widening the Range of Experiments on Subtle Energy Continues Deepening Our Understanding About the Nature of This Universal Phenomenon." Society for Scientific Exploration Annual Meeting, July 29, 2021.

Tiller, William; Dibble, Jr., Walter; and Kohane, Michael. *Conscious Acts of Creation: The Emergence of a New Physics*. Walnut Creek, CA: Pavior Publishing, 2001.

Chapter 11: Ra and the Law of One

Channeling by Jeanne Love and Regina Ochoa from the deceased astronauts of the space shuttle Challenger and the Columbia space shuttle. *https:// challengercc.org*

The Ra Contact: Teaching the Law of One, Volumes I and II, Carla Rueckert, channel. L/L Research. There is an Audible version of the two volumes. You also can download any and all of the book and channeling sessions at *https:// llresearch.org/home.aspx*

Chapter 12: On Spirituality and Science

Roberts, Jane, channel: *SETH Speaks: The Eternal Validity of the Soul*. New York: Bantam Books, 1980.

Schucman, Helen, channel: *A Course in Miracles (ACIM)*. 1976. *https://ACIM.org*

Jayem, channel: *The Way of Mastery*. Shanti Christo Foundation. *https:// wayofmastery.com/jeshua/*

List of Figures

Figure 1: A tiling with squares whose side lengths are successive Fibonacci numbers. Source: Images from Wikipedia: *https://en.wikipedia.org/wiki/Fibonacci_number*

Figure 2: The spiral is inherent in us and in nature. Source: Persaud-Sharma, D. and O'Leary, J.P. "Fibonacci Series, Golden Proportions, and the Human Biology." *Austin J Surg.* 2015;2(5): 1066.

Figure 3: Model of a torus. Source: Photograph by Jerry Gin of a torus sculpture purchased by Jerry Gin from Brian Berman at *https://www.bermansculpture.com/*

Figure 4: The key symbols of creation. Source: Created by Jerry Gin on PowerPoint; some component parts of symbols derived from internet, such as the image of Aum.

Figure 5: The Electromagnetic Spectrum as viewed by science today. Humans see only in a narrow band of the visible spectrum. Source: Wikimedia Commons, the free media repository, *https://upload.wikimedia.org/wikipedia/commons/c/cf/EM_Spectrum_Properties_edit.svg* Attribution: Inductiveload, NASA, CC BY-SA 3.0 *http://creativecommons.org/licenses/by-sa/3.0/>*, via Wikimedia Commons

Figure 6: Monochord and Tuning Forks: Through resonance, vibrational energy information is communicated, so that other strings or tuning forks at different octaves also vibrate. Source: *Back to a Future for Mankind: BioGeometry*, by Ibrahim Karim, Ph.D., Dr. Sc., published by BioGeometry Consulting, Ltd.

Figure 7: The string length or wavelength (W) of a monochord instrument is the same concept as the string length or wavelength (W) of a pendulum. Through resonance, the string length of the pendulum can detect vibrations of different octaves of the object being detected or measured. The strength of the vibration is the amplitude (A) and can be picked up by the diameter of rotation of the pendulum. Source: *Back to a Future for Mankind: BioGeometry*, by Ibrahim Karim, Ph.D., Dr. Sc., published by BioGeometry Consulting, Ltd.

Figure 8: The diagram on the left illustrates the distribution of 12 colors around a sphere based on the position of the sun. This represents the classification that radiesthesia uses to describe resonance with subtle energies. The diagram on the right shows the same distribution, but based on the compass direction on a circle. Negative green is not a real color; it represents the vibrations of the penetrating energy *opposite* green. Source: Diagram from *Back to a Future for*

Mankind: BioGeometry, by Ibrahim Karim, Ph.D., Dr. Sc., published by Bio-Geometry Consulting, Ltd.

Figure 9: Pendulums made from vials with an aspirin and with water. These pendulums will rotate clockwise when they are held above either water or aspirin pills, demonstrating your ability to detect resonance using a focusing device (the pendulum). Source: Photograph of experiment by Jerry Gin.

Figure 10: A pendulum made with a string and a nut. The nut is metal and the washer below the nut is metal. Because of resonance between the two similar metals, the pendulum swings clockwise once the to-and-fro motion has been initiated to overcome inertia. Source: Photograph of experiment by Jerry Gin.

Figure 11: The Wadj dowsing tool from ancient Egypt. Source: Image of Wadj seen in *https://medusa-art.com/egyptian-wadj-papyrus-column.html*

Figure 12: Pendulums made of salt, flour, water + string. Source: Photograph of pendulums made by Jerry Gin.

Figure 13: For practice, when you hold the neutral pendulum above a battery and initiate the to-and-fro movement, the pendulum rotates clockwise over the positive pole and counterclockwise over the negative pole. Source: Photograph by Jerry Gin.

Figure 14: Wavelength of the color red. Using a neutral pendulum, gradually lower the pendulum by sliding up the string with the fingers over the red surface. Overcome inertia by creating a to-and-fro movement. When the string length reaches the wavelength of red, the pendulum will rotate clockwise. The pendulum will not rotate clockwise over other colors but will do so over any other object that is red. Source: Photograph by Jerry Gin.

Figure 15: Pendulums made of acrylic (top) and wood (bottom). Source: Photograph by Jerry Gin.

Figure 16: How to find a pendulum's clockwise-rotation length (personal wavelength). Source: Photograph by Jerry Gin.

Figure 17: The BG16 pendulum (left) and the IKUP pendulum (right) are designed to easily and quickly detect BG3 through resonance with shape and numbers. The BG16 pendulum detects BG3, and the IKUP pendulum detects BG3 and can also be used as a neutral pendulum to find your personal wavelength. Both pendulums also emit BG3. Source: Picture of pendulums sold by BioGeometry distributors.

Figure 18: The above symbols have both BG3 and 3-6-9 subtle energy qualities, which can be detected by a BG16 pendulum and a 3-6-9 pendulum. Source: Symbols put together by Jerry Gin.

Figure 30: Dan Winter has found that compression waves that follow the golden ratio cause constructive interference, which results in other waves following a similar pattern (*phase conjugate*) to compress in the same manner. He defines the compression pattern as *implosion*. Spirals are activated by tracing them with a pencil. BG3 and 3-6-9 is found to be associated with the spiral curve. You can prove this by following the spiral with a BG16 or a 3-6-9 pendulum (the tools to detect BG3 and 3-6-9). Source: Dan Winter, *Fractal Conjugate Space & Time: Cause of Negentropy, Gravity and Perception: Conjuring Life: "The Fractal Shape of TIME" Geometric Origins of Biologic Negentropy...* (*http://www.fractalfield.com/conjugategravity*). BG3 and 3-6-9 energy observations and activation by Jerry Gin.

Figure 31: Schumann frequencies calculated from the golden ratio are very close to the frequencies measured by Otto Schumann. These frequencies are very similar to brain-wave frequencies, especially those found in the bliss state. The spirals are activated by being traced with a pencil to make it a continuous spiral. BG3 and 3-6-9 are found to be associated with the spirals. Source: *Fractal Conjugate Space & Time: Cause of Negentropy, Gravity and Perception: Conjuring Life: "The Fractal Shape of TIME" Geometric Origins of Biologic Negentropy...* (*http://www.fractalfield.com/conjugategravity*). BG3 and 3-6-9 energy observations and activation by Jerry Gin.

Figure 32: The Schumann frequencies—calculated from the golden ratio, when converted to units—are in resonance with BG3 and 3-6-9. The lines are activated by being traced with a pencil, so they can be considered a unit. Source: Drawn by Jerry Gin.

Figure 33: Golden ratio sequence shown as lengths reveals both BG3 and 3-6-9. Source: Drawn by Jerry Gin.

Figure 34: Tiling of successive Fibonacci numbers by creating squares of the numbers. Source: Wikipedia (*https://en.wikipedia.org/wiki/Fibonacci_number*).

Figure 35: The Fibonacci spiral created from the tiling of the squares from the Fibonacci sequence of numbers. Source: Wikipedia (*https://en.wikipedia.org/wiki/Fibonacci_number*).

Figure 36: "Phi in Ear and Hand." Source: "Fibonacci Series, Golden Proportions, and the Human Biology" by Dharam Persaud and James P. O'Leary (*Austin J Surg.* 2015;2(5): 1066.

Figure 37: Phi ratio in the human body and in the hand. Source: Drunvalo Melchizedek, *The Ancient Secret of the Flower of Life*, Volume 1. Flagstaff, AZ: Light Technology Publishing, 1999.

Figure 38: Symbols with 3-6-9 subtle energy qualities. For manifestation of clockwise (CW) rotations for BG3, the direct torus-related symbols need to be drawn in a CW manner for the "6" and "9" aspects of the torus and the 3-6-9 numbers. For Yin Yang, the center line must be traced to activate the symbol. For Rodin's symbol, the line connecting the numbers must be traced for activation. The Reiki Power symbol is normally activated, since it is drawn from the center in a clockwise motion; when drawn counterclockwise, it is not activated and shows negative BG3 (counterclockwise rotation of BG3 pendulum). Source: Created by Jerry Gin.

Figure 39: Schematic (not to scale) to show principle of BG3 ruler (also works for obtaining relative values of 3-6-9). Object to be tested is placed in the Witness circle, and a BG3 pendulum (BG16 or IKUP pendulum) is used to measure the relative level of BG3 or 3-6-9. Readings are made at each 90-degree line until no more BG3 is detected. Source: Drawn by Jerry Gin, derived from concepts by Ibrahim Karim in his courses on BioGeometry.

Figure 40: The table illustrates the relative BG3 and 3-6-9 values of the various symbols, as well as the nullifying effect of the BG3 and 3-6-9 values when the two types of symbols are put together. The last column shows the return of the subtle energy resonant values when a harmonizing yin yang symbol is added. Source: Drawn by Jerry Gin.

Figure 41: Aum placed in a 3-6-9 array. The 3 aum symbols are in the center, surrounded by 6 aums and finally 9 aums in the outer circle. This array produces BG3 and 3-6-9 of 6,400. Source: Drawn by Jerry Gin.

Figure 42: The 3-6-9 array of aum symbols was further reduced in size, and each array was placed in another 3-6-9 array, creating a fractal of the 3-6-9 arrays. This resulted in a BG3 and 3-6-9 level of 17,500. Source: Drawn by Jerry Gin.

Figure 43: The kinetics of water activation/structuring, as evidenced by the increase in BG3 levels when water was placed on top of a 3-6-9 array of a 3-6-9 aum array (fractal pattern in 3-6-9 geometry in above figure). In 96 hours, the BG3 of the water reached 7100. A control water was run at the same time, which did not show increase in BG3 and 3-6-9 values. Source: Jerry Gin.

Figure 44: Spiritual 12 aums surrounding the 3-6-9 array of aums resulted in a BG3 and 3-6-9 level of 13,500, as well as harmonization of the nullifying effect of 3-6-9 symbols on BG3 tools. Source: Drawn by Jerry Gin.

Figure 45: Fractal 3-6-9 Array of 3-6-9 aum array with 12 surrounding aums at each level. This produces the highest BG3 and 3-6-9 levels in the range of 135,000, completely off the scale of conventional BG3 rulers. Source: Drawn by Jerry Gin.

Figure 46: Activation/structuring of water using the Fractal 3-6-9 Aum Array with Spiritual 12 aums with each array, as shown in Figure 45. Source: Table by Jerry Gin.

Figure 47: Understanding the stages of nano materials. The figure summarizes the transition from solid matter to the monoatomic nano layer and the separation to the freed-up monatomic state called GANS. Both nano state and GANS show 3-6-9 subtle energy qualities. Source: From Keshe Foundation, as seen in *https://www.plasmaproduction.org/*

Figure 48: Formation of monatomic nano layers with use of a caustic environment. The close-up portion of the figure shows the gaps and MagGrav Fields—the center of energy and information exchange. Source: Keshe Foundation Spaceship Institute. *See https://kfssi.org/learn-more/* for series of 12 workshops.

Figure 49: Interaction of the fields of two magnets, or two GANS units, or monatomic units, or units in the nano layer. The gravitational/magnetic interactions dictate the gap or distance between the magnets or units comprising the MagGrav Field. Source: Keshe Foundation Spaceship Institute. *See https://kfssi.org/learn-more/* for series of 12 workshops.

Figure 50: The various states of matter and their energy density. The nano state is the fourth state of matter. It is a "CAPTURED" and CONCENTRATED GAS—in the magnetic-gravitational plasmatic state. Source: *http://www.plasmaproduction.org/*

Figure 51: Pictures from the Internet showing anomalous effects of Ormus on growth of a cat's tail that has been cut off, and the tremendous growth of fruits. Source: *https://cupdf.com/document/ormus-plant-comparison-pictures.html*

Figure 52: ER50 Brown's Gas Generator from Eagle Research. The unit shows a bubbler unit to generate Brown's Gas for inhalation. The other key accessory is a torch for burning Brown's Gas. Source: Brown's Gas Generator built by Jerry Gin from a kit made by George Wiseman. (See George Wiseman's kits and books at *www.eagle-research.com.*) Photo by Jerry Gin.

Figure 53: Effect of passing Brown's-gas flame on wood, copper wire, brass, and stainless steel. (A) Cube of wood lightly darkened with flame from Brown's gas. Wood now has 3-6-9 subtle energy. (B) Cube of wood lightly darkened with flame from propane torch. Wood does not show 3-6-9 subtle energy. (C) Copper wire heated white-hot with Brown's gas. Tip of wire was melted and entire wire shows blackened copper GANS, which has 3-6-9 subtle energy. (D) Brass tube lightly heated with Brown's gas shows 3-6-9 subtle energy. (E) Stainless steel lightly heated with Brown's gas shows 3-6-9 energies. Source: Photo and experiment by Jerry Gin.

Figure 61: To check the stability of the copied energies of 3-6-9 aum array of array, the energy was measured over 42 hours to see if the energy level remained stable. The data shows no deterioration of the energy level. Source: Table and experiment by Jerry Gin.

Figure 62: Kinetics of Water Activation with 3-6-9 Aum of Aum array Copied onto CD, Wood, Paper, and Stone. Source: Table and experiment by Jerry Gin.

Figure 63: Increase and decrease in pH, based on intention imprinted in the IHD. Source: Nisha Manek, *Bridging Science and Spirit: The Genius of William A. Tiller's Physics and the Promise of Information Medicine*. Conscious Creation, LLC, 2019.

Figure 64: Effect on reaction rate of alkaline phosphatase (ALP) upon exposure to IHD, with intention to increase reaction rate, as compared to UED with no intention. Source: Nisha Manek, *Bridging Science and Spirit: The Genius of William A. Tiller's Physics and the Promise of Information Medicine*. Conscious Creation, LLC, 2019.

Figure 65: Second Law of Thermodynamics. Source: Nisha Manek, *Bridging Science and Spirit: The Genius of William A. Tiller's Physics and the Promise of Information Medicine*. Conscious Creation, LLC, 2019.

Glossary

A

Array: A geometric arrangement of the symbols (e.g., bagua, aum, etc.). A square array, for example, would be 4 symbols at the corners of a square shape.

"As above, so below": God is the mind, the consciousness in everything. The underlying nature of everything is the same. All matter/creation is formed by light and possesses consciousness.

***Aum*:** A key symbol, chant, and ending word in prayers and meditations, in many religions. The Sanskrit aum is derived from 3-6-9. In Sanskrit, this word translates to Source or Supreme. In Buddhism, one explanation is that the "A" represents creation, the "U" is manifestation, and the "M" is destruction. This is the cycle of creation. It is the same in the Hindu religion. *Aum* represents the three aspects of God: the Brahma (A), the Vishnu (U), and the Shiva (M) aspects.

B

Bagua: An eight-sided symbol, containing the eight trigrams of the I Ching in Asian cultures. In Taoist cosmology, it represents the fundamental principles of reality. It may also represent the full lines and broken lines found in the sacred geometry of creation formed by the intersections of tetrahedrons.

Balance: I AM is at the fulcrum of the lights, balancing everything. Balance is found at the center. When everything is in balance, there is the kind of calmness and stillness that is achieved in deep meditation. When we deviate from the center, there is less balance, and therefore greater unhappiness/disharmony.

BG3 (or BioGeometry 3): BG3 is an energy quality that is found in sacred spots around the world, and in the centers of structures. Although it has three qualities that can be measured separately, it is actually a single quality of harmony and centering. This is important because the presence of the quality of BG3 *produces* harmony, balance, and centering. The three qualities found in BG3 are (1) horizontal negative green, (2) the higher harmonic of gold, and (3) the higher harmonic of ultraviolet.

BioGeometry: According to Ibrahim Karim, the founder of BioGeometry, the definition is: "The science of establishing harmony between biological fields and their environment, through the use of design, language of color, form, motion and sound."

Brown's Gas: The physical and chemical properties of Brown's gas are not fully established. One hypothesis is that it is monatomic hydrogen, while some other discussions suggest that it might be another form of water. It may be a combination of two monatomic hydrogens and one monatomic oxygen, interacting to form a new entity that has resemblance to water but is not water. The principle of the Browns-gas machine is electrolysis of water. Unlike normal electrolysis, the Brown's-gas machine has multiple metal plates between the anode and the cathode. The electrolyte solution connecting the electrodes is a solution of sodium hydroxide (lye). A gas is formed between the plates that is not molecular hydrogen and oxygen. The gas has GANS properties, since it has 3-6-9 subtle energy qualities.

C

Creation: In this half of the life cycle, the compression of the spiraling light takes place in a *clockwise* spiral. According to Walter Russell, the first half of the creation/decay cycle is generative and electric, compressing light into matter. (*See also* its opposite, "Decomposition" or "decay.")

D

Decomposition (decay): The second half of the life cycle, in which the unwinding) of the light in a *counterclockwise* spiral results in the radiative breakdown of matter. According to Walter Russell, the second half of the cycle is the breakdown or unwinding of matter and is radiative and magnetic. (*See also* its opposite, "Creation.") Scientists may term the decay/decomposition as "entropy."

E

Electronegativity: A measure of an element's ability to attract electrons.

Electropositivity: A measure of an element's ability to donate electrons.

EMF (electromagnetic fields): The field can be viewed as the combination of an *electric field* and a *magnetic field*. The electric field is produced by stationary charges, and the magnetic field by moving charges (currents); these two are often described as the sources of the field. In this book, what is discussed is the constant barrage of microwave electromagnetic fields (EMFs) from WiFi, cell towers, mobile phones, cordless phones, radar, Smart Meters, computers, phones, tablets, baby monitors, and many other devices that make up our modern high-tech world. Typical microwave frequencies range from 50 Hz at the extremely low range (wiring) to 300 GHz. ("Hertz" is the number of oscillations of the perpendicular electric and magnetic fields per second; a GHz is 109 Hertz.)

Entropy: The negative component to torsion energy that is involved with radiation/dissipation. The term means decomposition, as defined above—the breakdown of everything as it winds down. It is the second half of the creation/decay cycle of matter. (*See also* Negentropy.)

F

Fibonacci spiral: This is seen in the spirals of pine cones, the shape of sea shells, the branching nature of trees, and even the shape of a human being (all the segments of the body are in the Fibonacci spiral shape). In mathematics, each number of the Fibonacci sequence is formed as the sum of the two preceding ones, starting from 0 and 1.

G

GANS (Gas in Nano State): *Gas* in the GANS means a free-standing monatomic state. The *nano* layers are not independent but interact with the physical matter, with each other, and with the environment through magnetic and gravitational fields (MagGrav Fields). The term *nano* refers to one billionth of a meter. In Keshe's terminology, *nano* refers to entities at the nanoscale, such as atoms, which are independent structures. Other people have used the term *monatomic*. The atoms are independent structures and behave like round magnets on a surface. Anything can be converted to a GANS state. One of the primary GANS for this technology is CO2/ZnO from the copper- and zinc-plate method. However, other metals and non-metals also can be used to make GANS, such as calcium, food, seawater, soil, etc. (*See also* Monatomic and Ormus.) GANS, being MagGrav units, have 3-6-9 subtle energy. MagGrav structures resemble the structure of a torus, with energy flowing in and out as in a torus vortex.

GANS Field: The nano and GANS state results in formation of MagGrav Fields, or plasma fields, which in turn can lead to matter formation. Geometric arrays of containers containing GANS will form fields with measurable BG3 and 3-6-9 subtle energies.

Golden mean: 1.618. (*See* "Golden ratio," below.)

Golden ratio (Golden mean): The golden ratio (phi) is represented as a line divided into two segments "a" and "b," such that the entire line is to the longer "a" segment as the "a" segment is to the shorter "b" segment. In mathematics, two quantities are in the golden ratio if their ratio is the same as the ratio of their sum to the larger of the two quantities. Algebraically, it is (a+b)/a = a/b = phi. The ratio is the irrational number 1.618…. In Fibonacci numbers, the ratio of 2 sequential numbers approximate the golden ratio.

Gravitative (*see also* Radiative): In the context of this book, this represents the inward centripetal clockwise spiraling of light (vortex), resulting in the creation of matter. Its usage derives from the work of Walter Russell in his description of the creation of matter.

Grid lines: In addtion to the major ley lines, there also are grid lines covering the earth, both on the ground and above ground. The grid lines carry information, which can be either beneficial or detrimental, depending on the direction of rotation of upward spiraling energies emanating from crossing associated with the lines. The *Hartmann lines* run in a north-south and east-west direction. The *Curry lines* run diagonal to the Hartmann lines. The Benker Cube System is layered above the Hartmann grid system. (*See also* Ley lines.)

L

Ley lines: Earth ley lines are earth grid lines linking sacred power spots. They were located using vibratory rods or pendulums. (*See also* Grid lines.)

M

MagGrav: This is a term coined by Mehran Keshe to describe the magnetic and gravitational nature of GANS, Nano, and Plasma. The individual units of this fourth state of matter have an inward flow (gravitational) and an outward flow (magnetic). Magnets may be thought of in this manner. There is a north and south pole of the magnet, which may be better thought of as "MagGrav" with its magnetic (repulsion) and gravitational (attraction) aspects.

MagGrav Fields: MagGrav units such as GANS will induce other MagGrav units to space themselves relative to each other because of their magnetic and gravitational properties. This is analogous to putting disc magnets on a table and their adjusting themselves because of their attractive-repulsive properties. Creating a geometric array of vials of GANS will induce formation of BG3 in the field by balancing the MagGrav unit within the field.

MagGrav torus units: The torus structure is a good way to think about GANS, Nano, and Plasma. In matter formation, a torus is the vortex in light (gravitational) and out of light (magnetic), which is analogous to the definition of GANS.

Matter: Using the Walter Russell definitions, creation of matter is based on the spiraling in of light in a centripetal clockwise vortex motion. Decay of matter is the spiraling out of light in counterclockwise vortex motion.

Monatomic: Individual units of atoms not bound to other atoms. Mehran Keshe describes the fourth state of matter as GANS and Nano. Ormus is one form of GANS.

Motion-in-Equilibrium: In the Walter Russell descriptions of the formation of matter, matter forms based on the vortex of two spirals of light meeting at the apex. The two spirals, rotating very rapidly, reach a "motion-in-equilibrium" to give the semblance of matter before progressing to spiral into the decay portion of their cycle.

N

Nano: One billionth of a meter (a sheet of paper is 100,000 thick). In Mehran Keshe's terminology, *nano* refers to entities at the nanoscale, such as atoms, which are independent structures. Others have used the term *monatomic*, since the atoms are independent structures.

Negentropy: The spiral clockwise direction of growth, or matter creation. For everything that dissipates due to entropy, there is a corresponding generative/creative effect that occurs. In classical thermodynamics, the universe only winds down, and is called entropy. More recent thinking is that entropy is balanced by negentropy, the formation of life and matter. Another term people now use is "syntropy," which has the same meaning as negentropy. (*See also* Entropy.)

O

Odic Energy (OD): The name given to the vital energetic force investigated by German scientist Baron Carl von Reichenbach (1788–1869), after the Norse god Odin. Today, we would call this "subtle energy."

Orgone energy: A term used describe the energy being radiated from the bions in experiments achieved by Dr. Wilhelm Reich (1897–1957), an Austrian doctor of medicine and a psychoanalyst. Orgone accumulates when insulators and metal are put together. Reich was also responsible for the orgone accumulator (based on the observation that orgone is ubiquitous in the atmosphere and other spaces), and its use as a cloud buster to create rain. Orgone has characteristics of subtle energy, and has the subtle energy quality of 3-6-9, implying the presence of GANS or MagGrav units.

Orgonite: Material often made from plastic (such as polyester resin) containing metal pieces and some fragments of crystals (e.g., quartz), which acts as a natural accumulator of orgone.

Ormus: The first GANS to be discovered. Initially, it was called Orbitally Re-arranged Monatomic Elements, or ORMEs. The science of Ormus (and GANS) is in its infancy. There are amazing stories of the biological effects of Ormus (e.g., a cat growing back its tail, oranges growing to the size of grapefruits, etc.). It is one of the easiest GANS to make, and is useful for creating BG3 and 3-6-9 fields. (*See also* GANS.)

Out of Body Experience (OBE): OBE is the phenomenon of being out of the body, and is sometimes called "astral travel" or "projection." The initiation of the experience often happens in the hypnogogic state—a state between being awake and before falling asleep. In that state, different dimensions/realms can be visited. Organizations such as IAC (International Academy of Consciousness) and the Monroe Institute teach how to have an out of body experience.

P

Pendulum: in this book, the pendulum is a tool that can be used to come into resonance with subtle or vibrational energies. Everything is in a state of vibration, so virtually everything can be detected and measured, based on coming into resonance with what you are trying to detect/measure. Just as a musical note is determined by the string length on a monochord instrument and causes all notes higher or lower in different octaves to vibrate, so the string length of a pendulum will cause a pendulum to rotate clockwise when it is in resonance with the energy that is to be measured. The pendulum is the focusing tool for detecting the resonance, which your right brain (or subconscious mind) can detect and transmit the information to the pendulum in an unconscious manner.

Personal wavelength: When you hold the string of a pendulum where it meets the pendulum body, and you gradually release the pendulum body and thereby extend the length of the string above the back of your hand, *the string length at which the pendulum starts rotating clockwise is called your "personal wavelength."* It is the wavelength of you, and has many useful applications described in the book.

Plasma and Plasma Fields *(according to Mehran Keshe)*: "Every physical object (electron, atom, grain of sand, plant, animal, human, planet, star, galaxy, universe, etc.) is nothing more than an accumulation of magnetic fields which Keshe refers to as *plasmatic magnetic fields*. These plasmatic magnetic fields consist of a huge quantity of individual magnetic fields. Plasmatic magnetic fields are three-dimensional, the movement and area of which is determined by the individual magnetic fields within the object. Each of these plasmatic objects is referred to as a *plasma* and is defined by the field content within it.

"The Keshe Foundation does not use the word 'plasma' in the same way that it is used in standard physics. The Foundation does not refer to the state of an ionized gas when they say 'plasma.' Plasma is defined by the Foundation as the entire content of fields which accumulate and create an object. It is NOT defined by its physical characteristics like ionization and temperature, for example. Plasma refers exclusively to the properties of the fields of which an object consists.

"In most cases, fields extend far beyond the physical boundaries of the object in question. As well, there are field compositions (plasmas) that are already detached from their physical source and are moving in a non-physical form (as a pure field body) through space to another physical (or non-physical) object. The movement of these field conglomerations is considered to be a kind of 'flowing.' The Foundation therefore defines energy as fields that move and flow in space from point A to point B." *https://en.kfwiki.org/index.php/Category:Introduction_to_Plasma_Science*

R

Radiative: In this book, "radiative" is derived from the terms used by Walter Russell to describe the decay potion of the creation/decay cycle of matter. Light is wound up in a centripetal spiral to create matter (gravitative), and is unwound in its decay cycle (radiative). (*See also* Gravitative.)

Radiesthesia: According to founder Dr. Ibrahim Karim, radiesthesia is "The science of using the vibrational fields of the human body to access information about other objects of animate or inanimate nature by establishing resonance with their energy fields, using specially calibrated instruments and a scale of qualitative measurement to decode this information."

Reiki power symbol: This symbol has both BG3 and 3-6-9, and shows the direction of rotation of creation within the torus.

Resonance: Resonance occurs when there are vibrations, and those vibrations induce other vibrations at different octaves. On a pendulum, by varying the string length, one can find a length that will be in resonance what one is trying to detect or measure.

S

Schumann resonance frequencies: Otto Schumann discovered that there are energy peaks in the spectrum of extremely low frequencies (ELF) that are the effect of global electromagnetic resonances which have been excited by lightning discharges. One might consider that the Schumann resonance frequencies are akin to the frequency of the Earth. Humans/animals require these frequencies for health. Astronauts deprived of these frequencies are negatively affected. These frequencies are around 7.83 Hz (fundamental), 14.3, 20.8, 27.3, and 33.8 Hz

SETH—Spiritual Energy for the Transformation of Humanity *(oversoul)*: The channeling of SETH is described in this book.

Spirituality, new: The new spirituality is knowing who we are—the I AM.

Syntropy, Law of (*also called* **Law of Negentropy):** A fourth law of thermodynamics proposed by Dr. Klaus Volkamer, to explain how subtle matter can counteract the second law of thermodynamics (the law of increasing entropy in spontaneous processes of gross matter).

T

Thermodynamics, second law of: This law is concerned with the concept of Free Energy and Entropy. It is the free energy that allows work to be done. Entropy is a component of the equation for free energy. By decreasing entropy, work increases. Life and intention—which can translate to information—would increase free energy by decreasing entropy.

3-6-9: A subtle energy quality associated with the torus structure. Creation of matter is associated with the torus. Thus, structures, symbols, and numbers associated with creation and the creation process have the 3-6-9 energy. This subtle energy is detected by resonance with the 369 elements of the torus (described in the book). The elements of 369 can be drawn or inscribed into a pendulum so that the pendulum is in resonance with any 3-6-9 subtle energies. GANS and plasma, being elemental torus structures, have the 3-6-9 energies.

Torsion energy: Dr. Nikolai Kozyrev coined the term "torsion energy." He found that there is an energy which is counter to entropy, the breakdown of matter. His experiments proved the existence of torsion energy. Today, we might call this a form of subtle energy.

Torus: The torus is a doughnut-shaped whirlpool vortex that is "the only manner by which self-sustained motion can exist in a given medium." (Arthur Young, *The Reflexive Universe,* p. xxi.) The shape is associated with the electro-magnetic field of an atom, a seed, a human, a planet, a solar system, a galaxy, the Universe. In this book, the basis of matter creation is light spiraling centripetally inward in a vortex to form matter, with the outward spiral resulting in the disintegration of matter. These inward and outward spiraling lights describe the formation of the torus.

V

VELO Technique (Voluntary Energetic Longitudinal Oscillation): An energy practice that helps you learn the feel of energy in the body. The International Academy of Consciousness teaches this technique as part of the practice in learning about the out-of-body experience.

Vibrational level, higher: "Higher vibrational level" means that the revolution of light in the spiral is closer to the apex or center, where light revolves the fastest. Your vibrational level goes up when you are stating affirmations that bring you to your I AM essence.

W

Wadj: A pendulum tool used in ancient Egypt for measuring as well as giving off subtle energies.

Wavelength: For a pendulum, the wavelength is the length of the string between the index finger and thumb, and the weight on the pendulum.

Witness: A Witness is a representation of something. In remote healing, it could be a picture of a person to be healed. In radiesthesia, a witness (a picture or cells from wiping one's forehead) is used to represent the person to test if a remedy is effective or not effective for that person. On a BG3 Ruler, it is a space (circle) for placing something you want to test for its level of BG3 (see Chapter 9).

Y

Yin yang: A symbol that can be used to bring in harmony—such as with our yin and yang natures. From our testing and observations, the structure is derived from the torus, with the dividing curve of the circle having both clockwise and counterclockwise directions representing the two halves of creation-disintegration.

About the Author

Jerry Gin is the chairman and CEO of FMBR (Foundation for Mind-Being Research), a non-profit organization that seeks to advance the consciousness of individuals and organizations to enable us to live in greater harmony with one another, the earth, and the cosmos. His credentials span both science and business, with a PhD in Biochemistry from UC Berkeley, a BS from the University of Arizona, and an MBA from Loyola College. He has held Director positions at Dow Pharmaceuticals and Syva/Syntex. His professional science activities include laboratory medicine, clinical chemistry, pharmaceuticals, ophthalmology, and dentistry. His entrepreneurial activities include founding ChemTrak (developed the first home cholesterol test), Oculex Pharmaceuticals (developed the first intraocular drug delivery), Nuvora (developed products that offer sustained delivery for the mouth), Visionex (ophthalmic diagnostic), and Livionex (dental gel, metal modulation). Jerry's driving passion for the past 20 years has been exploring the nature of the universe. To this end, his activities have included research, studies in many diverse disciplines, experiential activities in self-exploration, and growth and knowledge through inner knowing. His first book in this series, *The Seeker and The Teacher of Light*, was a culmination of his views as to the nature of the universe, as well as the views of his friend, Joachim Wippich. Jerry's website is *https://jerrygin.com/*. He can be reached at *jerry@jerrygin.com*.

CPSIA information can be obtained
at www.ICGtesting.com
Printed in the USA
LVHW040344020622
720206LV00001B/28